Russia's Limit of Advance

Scenarios

BEN CONNABLE, ABBY DOLL, ALYSSA DEMUS, DARA MASSICOT,
CLINT REACH, ANTHONY ATLER, WILLIAM MACKENZIE,
MATTHEW POVLOCK, LAUREN SKRABALA

Prepared for the United States Army
Approved for public release; distribution unlimited

For more information on this publication, visit www.rand.org/t/RR2563z1

Library of Congress Cataloging-in-Publication Data is available for this publication.
ISBN: 978-1-9774-0244-8

Published by the RAND Corporation, Santa Monica, Calif.

Preface

This report documents the scenarios developed to support the research and analysis presented in the RAND report *Russia's Limit of Advance: Analysis of Russian Ground Force Deployment Capabilities and Limitations*, available online at www.rand.org/pubs/research_reports/RR2563. The two reports were produced as part of the project *Defeating Russian Deployed Joint Forces*, sponsored by the Office of the Deputy Chief of Staff, G-3/5/7, U.S. Army. The purpose of the project was to assess challenges that deployed Russian forces pose to U.S. Army forces; identify opportunities to defeat Russian deployed forces in a range of environments and at various levels of conflict; identify limitations to Russia's ground force deployment capabilities, including logistics, lines of communication, deployed force protection, air defense, system ranges, command and control, and joint integration; and recommend ways for the U.S. Army and the joint force to defeat Russia's deployed forces in multiple prospective combat scenarios.

This research was conducted within the RAND Arroyo Center's Strategy, Doctrine, and Resources Program. RAND Arroyo Center, part of the RAND Corporation, is a federally funded research and development center (FFRDC) sponsored by the United States Army.

RAND operates under a "Federal-Wide Assurance" (FWA00003425) and complies with the *Code of Federal Regulations for the Protection of Human Subjects Under United States Law* (45 CFR 46), also known as "the Common Rule," as well as with the implementation guidance set forth in U.S. Department of Defense (DoD) Instruction 3216.02. As applicable, this compliance includes reviews and approvals by RAND's Institutional Review Board (the Human Subjects Protection Committee) and by the U.S. Army. The views of sources utilized in this study are solely their own and do not represent the official policy or position of DoD or the U.S. government.

Contents

Figures and Table

Figures

Table

Summary

By the time of its 2014 incursion into Crimea, Ukraine, Russia had regained a significant portion of the military power it lost after the fall of the Soviet Union, reemerging as a perceived threat to democracy. It soon became clear that Russia had broader interests than Europe—and perhaps a capacity to realize wider-ranging military objectives. Since the mid-2000s, Russia has been quietly accelerating its global engagements and has, more recently, increased its interests in Venezuela, various African states, and Asia. These developments have spurred renewed interest in Russian capabilities in the analytic community.

The focus of this research, Russia's ground combat deployment capability, stemmed primarily from sponsor requirements and resource limitations, but the insights from this analysis help fill an important knowledge gap that extends beyond an understanding of Russia's ability to support ground deployments. We argue that the capacity to deploy ground combat units is a better measure of overall conventional power projection than air or naval power alone. Air and naval forces are limited by an array of overflight and passage restrictions, but they also benefit from international agreements that guarantee considerable freedom of movement. In contrast, ground deployment depends on and reflects global and regional diplomatic influence or, alternatively, brute force to obtain on-the-ground access. Air and naval forces can be deployed independently, but ground forces require joint and, often, combined operations that tax a broader cross-section of the Russian military infrastructure.

This report presents notional Russian Ground Force (RGF) military deployment scenarios that informed the analysis in a companion report, *Russia's Limit of Advance: Analysis of Russian Ground Force Deployment Capabilities and Limitations* (available at www.rand.org/pubs/research_reports/RR2563). That analysis examined seven notional scenarios, using one deployment to illustrate the analytical process: the Kuril Islands. This report presents detail on the five other scenarios that we analyzed to generate the findings presented in that report, as well as an additional informative scenario on Ukraine (our "+1" scenario). Table S.1 summarizes all seven scenarios.

Each chapter of this report is dedicated to one of six scenarios (excluding the Kuril Islands) and includes slides from a series of larger briefings prepared for this project. We selected slides that were particularly relevant to the focus of our analysis—RGF deployment capability. In the interest of brevity, we do not include informational slides about the scenarios. However, each chapter opens with a brief overview of the scenario it addresses.

Table S.1
Summary Scenario Descriptions

Location	Description	Range
Kazakhstan	Russia and China engage in conventional combat in Kazakhstan	Border
Kuril Islands	Russia deploys to repel Japanese forces, conventional combat	Near
Tajikistan	Islamic State threat spills over into Tajikistan, Russia deploys to defend	Near
Serbia	Deployment to help put down an anti-government revolt in Serbia	Far
Syria	Rescue of surrounded Spetsnaz and Syrian military forces at Palmyra	Far
Venezuela	Stability operation in support of the Venezuelan government	Far
Ukraine	*Seizure of parts of Ukraine for incorporation into the Russian state*	*Border*

NOTE: The Kuril Islands scenario is not included in the chapters that follow because it is covered in detail in the accompanying report, *Russia's Limit of Advance: Analysis of Russian Ground Force Deployment Capabilities and Limitations*, Santa Monica, Calif.: RAND Corporation, RR-2563-A, 2019. The Ukraine scenario involved too many forces to allow precise analysis. However, we included it as an additional, informative scenario. Thus, we refer to it as a "+1" case in this report.

Caveats

These scenarios are strictly notional. The purpose of developing and presenting the scenarios was to explore various permutations of Russian ground combat power deployment capability, not to explore politically viable national security scenarios. The scenarios do not forecast any particular political events, nor should they be interpreted as presenting conclusions about Russian combat capabilities. In fact, we chose the scenarios with the knowledge that they might have limited political feasibility.

All information that we used to develop the scenarios is drawn from open sources; the bibliography at the end of this report lists the materials that we consulted, grouped by topic. See the companion report for our full analysis and findings.

Acknowledgments

We thank MG William Hix for sponsoring our research. MG Christopher McPadden supported the continuation and completion of this project through 2018. Our project monitor, LTC Andrew Brown, also provided valuable support, feedback, and insights throughout the research process. RAND Arroyo Center staff, including Strategy, Doctrine, and Resources program director Sally Sleeper and Francisco Walter, were instrumental in creating this research opportunity and in supporting our efforts. We also thank our Army sponsor staff, including Tony Vanderbeek and Mark Calvo, for their continuing interest in our research and for supporting our work with enthusiasm.

We are grateful to RAND colleagues Raphael Cohen and Ryan Schwankhart and to our external reviewer, Kimberly J. Marten, chair of the Department of Political Science at Barnard College, all of whom provided insightful reviews and feedback that helped shape this report and its companion volume.

Abbreviations

APC	armored personnel carrier
APOD	aerial port of debarkation
APOE	aerial port of embarkation
BTG	battalion tactical group
IFV	infantry fighting vehicle
MR	motorized rifle
MRAP	mine-resistant, ambush-protected (vehicle)
MRL	multiple rocket launcher
MTO	motor transport operation
RGF	Russian Ground Forces
SAM	surface-to-air missile
SOF	special operations forces
TAA	tactical assembly area
UAV	unmanned aerial vehicle
VDV	Vozdushno-Desantnye Voyska [Russian Airborne Forces]

CHAPTER ONE

Kazakhstan Scenario

In the Kazakhstan scenario, Russia deploys ground combat forces to Kazakhstan to counter Chinese intervention and protect Russian civilians. This is a border case involving the deployment of almost 14,000 troops. The purposes of this scenario were to test Russia's deployment capability in a location with clear trade-offs between rail and airborne movement and to show how even a scenario just outside Russia's Western and Southern military districts can be challenging.

Figures 1.1–1.7 show, respectively, the deployment range, available Russian forces, movement plan, initial-wave assumptions, second-wave assumptions, and ground movement assumptions.

Figure 1.1
Kazakhstan Scenario Deployment Range

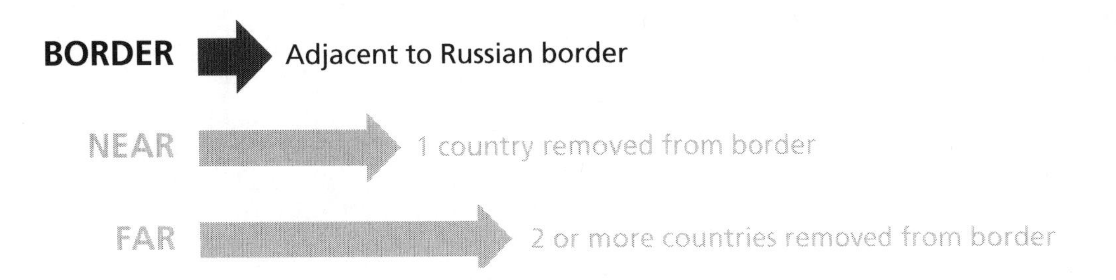

Figure 1.2
Russian Forces in the Kazakhstan Scenario

Joint Task Force Command Headquarters
Yekaterinburg

GROUND FORCES

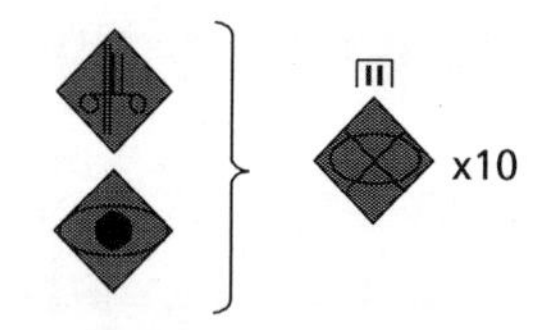

Mechanized BTGs
- 328 BMP-series IFVs
- 166 BTR-series APCs
- 40 MT-LB APCs
- 168 T-72s

Artillery group task-organized with BTGs
- 40 BM-21s
- 48 2S19s
- 32 2S3s
- 8 2S34s
- 32 Shturm-S systems

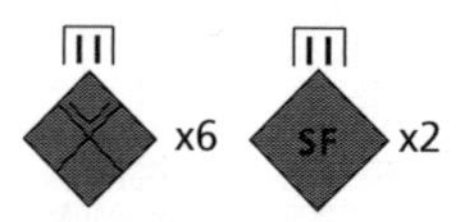

VDF (airborne) and Spetsnaz
- 75 BMP-series IFVs
- 78 BMDs
- 136 BTR-series APCs
- 24 Tigrs (light jeeps)

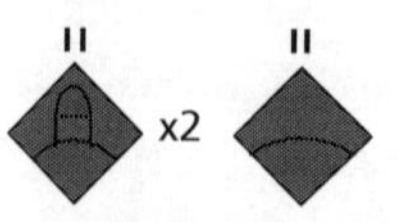

Air defense
- 6 Buk missile systems
- 16 S-300/SA-10s

Logistics support
1,632 vehicles

AIR

- 1 squadron helocopter rescue; 18 Mi-8 and Mi-26s
- 1 transport regiment; 16 IL-76 Candids
- 1 flight by SU-24MR reconnaissance aircraft
- 1 flight by SU-25SM fighter aircraft
- 2 regiments of SU-27s, SU-30s on combat alert

Total personnel	13,820 not including air-naval
Combat vehicles	1,562
Support vehicles	2,095
Combat aviation	90

Justification: Provide rapid-reaction force capable of extraction while also deterring Chinese military incursion into Kazakhstan

NOTE: APC = armored personnel carrier. BTG = battalion tactical group. IFV = infantry fighting vehicle. VDV = Vozdushno-Desantnye Voyska [Russian Airborne Forces]. Some vehicle-type abbreviations in this and similar figures, such as MT, BMD, BMP, BRDM, and BTR, are transliterated acronyms commonly used by the U.S. defense analytic community. For example, BMD is *boyevaya mashina desanta*, or *airborne combat vehicle*.

Figure 1.3
Kazakhstan Scenario Movement Plan

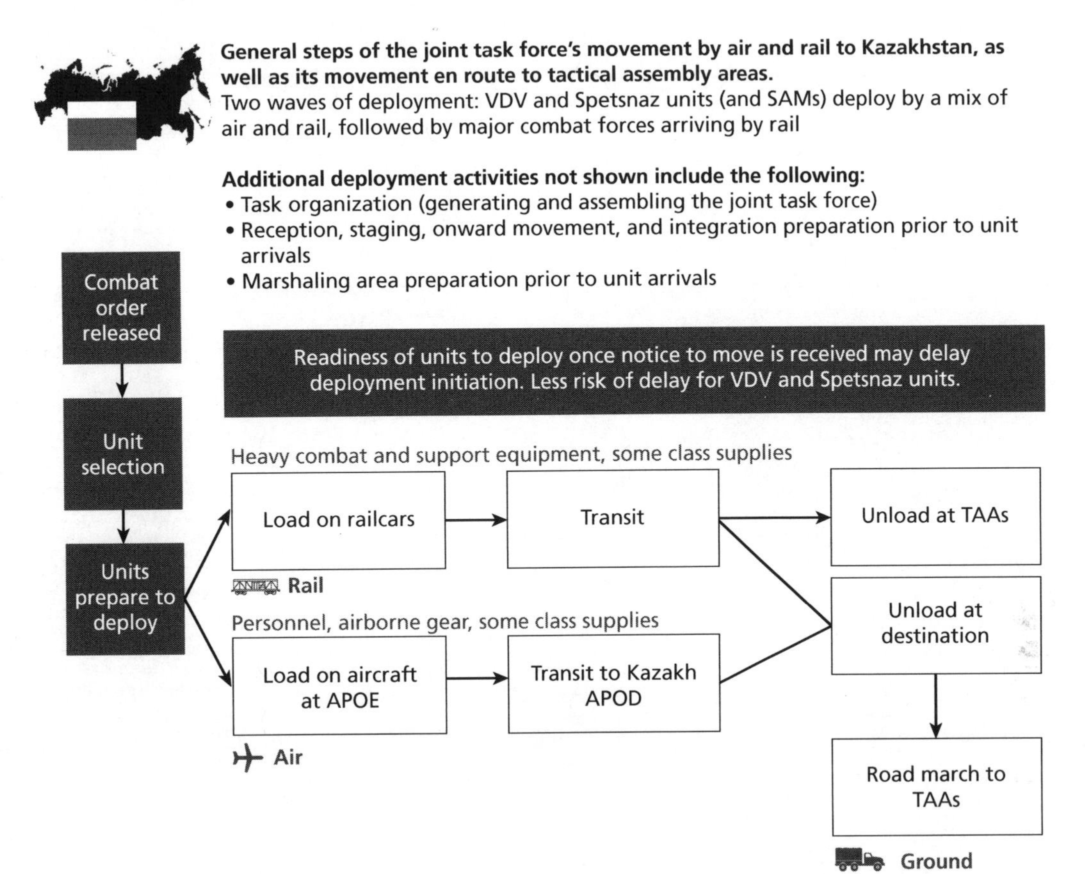

NOTE: SAM = surface-to-air missile. APOD = aerial point of debarkation. APOE = aerial point of embarkation. TAA = tactical assembly area.

Figure 1.4
Kazakhstan Scenario Initial Deployment Wave Assumptions

Assumptions
- Personnel, equipment, and some class supplies for VDV, Spetsnaz, and one SAM unit deploy to Aktau and Astana.
- 60 of 110 Il-76s and 6 of 9 An-124s are available.

Demand to lift initial force package exceeds available airlift inventory, necessitating two rounds of transport with a portion of the aircraft needing to make roundtrips.
With the extra turnaround time required, it would take at least 7.5 days to close the equipment and personnel. This does not include the airlift of class supplies, whose inclusion would further delay closure.

Airlifting only the Aktau force package (~3 days to close) and then railing the Astana force package (151 railcars, 5 trains) leads to closure of total initial wave deployment in around 4–5 days.

	Assess to lift	An-124	Il-76	% of equipment lifted	Personnel lift requirement	Origin	Destination
Wave 1	24th Spetsnaz Brigade	4	0	100	3 large aircraft	Novosibirsk	Aktau
	56th Air Assault Brigade	2	36	100	12 large aircraft	Kamyshin	Aktau
	10th Spetsnaz Brigade	0	14	100	8 large aircraft	Molkino	Astana
	31st Air Assault Brigade	0	10	23	12 large aircraft	Ulyanovsk	Astana
	Sortie totals	6	60	NA	35 aircraft	NA	NA

	Assess to lift	An-124	Il-76	% of equipment lifted	Personnel lift requirement	Origin	Destination
Wave 2	31st Air Assault Brigade	6	19	77	0	Ulyanovsk	Astana
	11th Air Assault Brigade	0	7	100	12 large aircraft	Sosnovy Bor	Astana
	297th Anti-Aircraft Missile Brigade	0	14	100	1 large aircraft	Alkino	Astana
	Sortie totals	6	40	NA	13 aircraft	NA	NA

NOTE: Calculations do not take into account the airlift of class supplies.

Figure 1.5
Kazakhstan Scenario Second Deployment Wave Assumptions

Assumptions

- MR and SAM units move equipment by rail to Aktau and three tactical assembly areas in eastern Kazakhstan: Oskemen, Aktogay, and Taldykorgan.
- Railcars and trains are readily available and in position when units are ready to deploy.
- 1 day at origin rail loading point, travel speed of 40 km/hr.
- Destinations clear train load every 4 hours.

Assets to lift	Total railcars (equipment + personnel)	Number of trains	Origin	Destination	Destination (km)	Time (days to load+ travel+ unload
511th Guards SAM Regiment	26 (22 + 4)	1	Engels Air Base	Taldykorgan	3,125	4.4
185th SAM Regiment	24 (20 + 4)	1	Yekaterinburg	Oskemen	2,037	3.3
74th Guards MR Brigade	215 (185 + 30)	4	Yurga	Aktogay	1,119	2.9
35th MR Brigade and 106th MTO Brigade	683 (593 + 45)	12	Aleysk	Taldykorgan	937	4.0
37th MR Brigade	211 (181 + 30)	4	Khyagt	Oskemen	2,757	4.5
15th MR Brigade and 105th MTO Brigade	604 (559 + 45)	11	Roshchinsky	Aktau	1,983	4.9
21st MR Brigade	215 (185 + 30)	4	Totskoye	Aktau	1,582	3.3

NOTE: Railcar calculations do not include class supplies. This could at least double the railcar demand.
MR = motorized rifle. MTO = motor transport operation.

Figure 1.6
Kazakhstan Scenario Ground Movement Assumptions

Assumptions
- There is unopposed, faster administrative ground movement to TAAs.
- There are preestablished forward logistics areas.
- Spetsnaz remain in Astana.

TAA	Almaty	Taraz
Unit	31st Air Assault Brigade	11th Air Assault Brigade
Distance (km)	1,215	1,601
Vehicles	196	45
Column length (km)	11.8 (day) 17.6 (night)	2.7 (day) 4.0 (night)
Completion time (days)	2.0	2.6
Assumptions	• 50 m vehicle spacing during the day • 75 m spacing at night • 30% time spent stopped for rest/maintenance • Road conditions will limit movement to an average of 40 km/h during the day, 30 km/hr at night	

Delays may be caused by weather, terrain, movement at night, or vehicle breakdowns.

NOTE: Estimates assume formation along a single road. Other options are stagger or diamond formations if road width allows.

Key Points from the Kazakhstan Scenario

Russia could deploy its ground force by air, rail, or road relatively quickly, absent Chinese intervention or transportation failures and without considering the class of supply movement (e.g., fuel, food, water, ammunition). Adding sustainment requirements and assuming even noncombat disruption would set the above timelines back days, if not longer. This movement is also highly vulnerable to combat disruption. Chinese interference with the limited road and rail networks or even minimal interference with the airfields—say, a cyberattack against air traffic control or a special operations raid against airfield support teams—could put Russia in an untenable situation.

CHAPTER TWO

Tajikistan Scenario

In this scenario, an extremist group similar to the Islamic State expands its operations into Tajikistan, threatening Russian bases and interests there. Russia deploys a ground combat force to secure its facilities and personnel, as well as to disrupt the group's activities in Afghanistan with fires and raids.

Figures 2.1–2.6 show, respectively, the deployment range, available Russian forces, movement plan, airlift assumptions, rail assumptions, and movement to tactical assembly areas.

Figure 2.1
Tajikistan Scenario Deployment Range

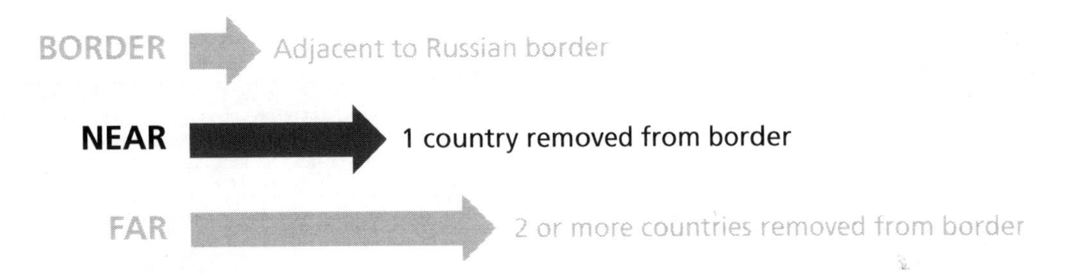

Figure 2.2
Russian Forces in the Tajikistan Scenario

GROUND FORCES

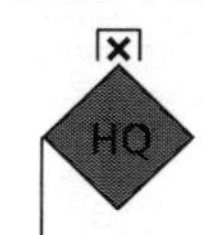

Joint Task Force Command
41st Combined Arms Army

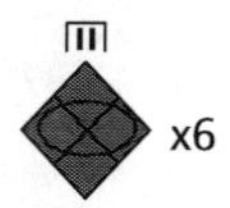

6 MR BTGs with
- 248 BMP-2 IFVs
- 48 BTR-80 APCs
- 22 MT-LB APCs
- 6 BRDM-2 reconnaissance vehicles
- 84 T-72B3 tanks
- 28 T-72BM tanks

BORDER TROOPS

4 detachments with
- 320 BTR-80 APCs
- 60 Ural-4320 trucks
- 24 2S1 SP howitzers
- 24 2S12 Sani mortars
- 24 2S9 Nona mortars

Spetsnaz
- 1 battalion (25) BTR-80 APCs
- 12 Tigrs (light jeeps)

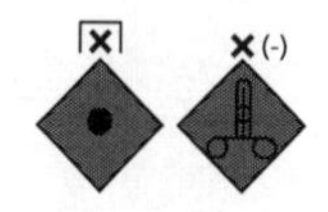

Artillery Group
1 artillery brigade, 0.5 missile brigades
- 28 BM-21 MRLs
- 24 2S3 Akatsiya self-propelled artillery
- 66 2S19 Msta-S howitzers
- 8 2S34 Khosta-S howitzers
- 24 Sani mortars
- 8 Uragan MRLs
- 6 Iskander-M transporter-erector-launchers

AIR

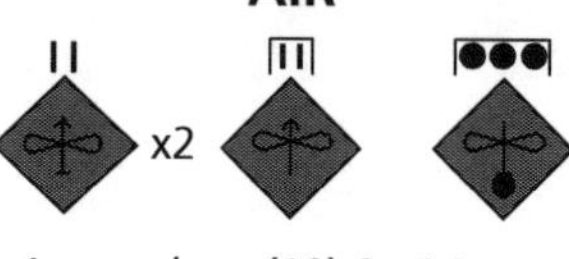

- 1 squadron (10) Su-34 attack aircraft + support
- 1 squadron (12) Mi-24P helicopters + support
- 1 squadron (12) Mi8AMTSh helicopters + support
- 3 Tu-22M3 bomber aircraft
- 1 squadron (11) Su-25 attack aircraft
- 2 Mi-8 helicopters
- 2 An-26 transport aircraft

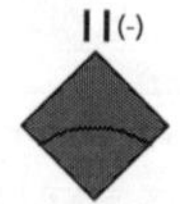

Air Defense
3 anti-aircraft battalions (-)
- 13 2S6M Tunguskas
- 54 9K38 Igla man-portable air defense systems

SUPPORT

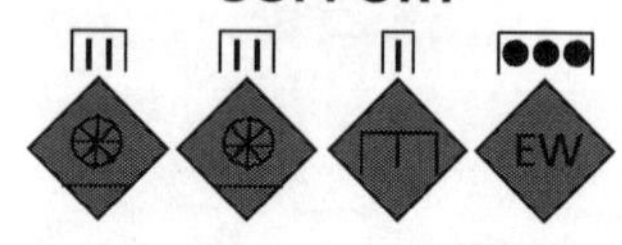

- 1 Leer-3 electronic warfare system
- 78 Zastava UAVs
- 12 Granat-1 UAVs
- 15 Orlan-10 UAVs
- Support vehicles

Total ground personnel	12,140
Combat vehicles	1,213
Support vehicles	1,193
Rotary aviation	26
UAVs	105

Justification: Deploy a self-sustaining joint combat team capable of reconnaissance-weapon and reconnaissance-strike counterterrorism operations and border security

NOTE: MRL = multiple rocket launcher. UAV = unmanned aerial vehicle.

Figure 2.3
Tajikistan Scenario Movement Plan

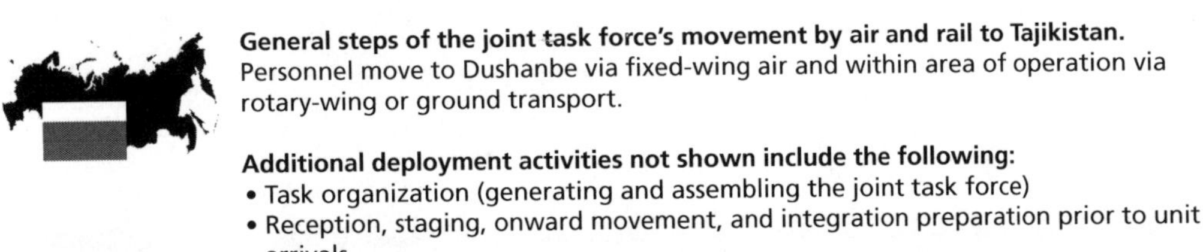

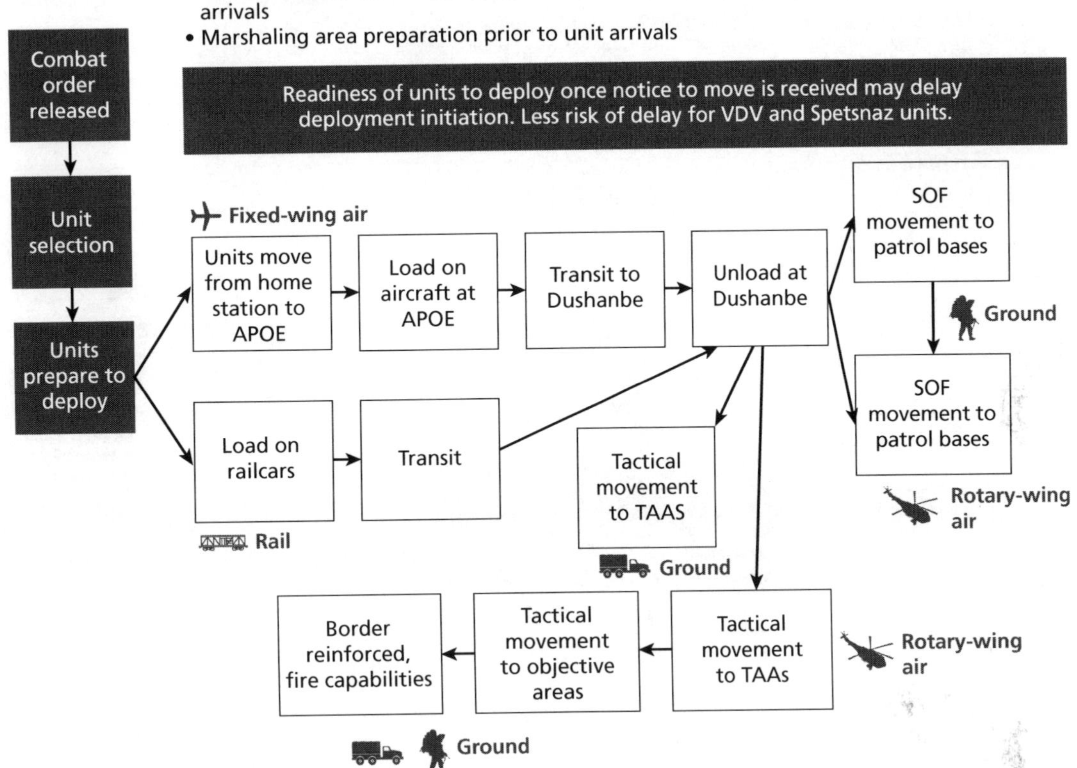

NOTE: SOF = special operations forces.

Figure 2.4
Tajikistan Scenario Airlift Assumptions

Assumptions
Spetsnaz and rotary-wing units' equipment and personnel deploy by air.

Assess to lift	Lift requirement (equipment + supplies)
Spetsnaz units	7–9 Il-76s or 3 An-124s
Rotary-wing units	6 An-124s to deploy Mi-24s 6 Il-76s or 4 An-124s to deploy Mi-8s
Sortie totals	13–16 Il-76s or 9–13 An-124s
% of estimated available fleet*	~25% of Il-76s 150–217% of An-124s

*Assumes 60 of 110 Il-76s and 6 of 9 An-124s are available.

1 leg; 2,121 km; 3 hours

Novosibirsk
Новосибирск
UNCC
Astana
Астана
Kazakhstan
Uzbekistan
Kyrgyzstan
DYU

Calculations for fixed-wing movement only.

Possible deviations from "best-case" air deployment

Risk
Adequate airlift is not available to deploy helicopters.

Mitigating option: Self-deploy
- This increases maintenance issues.
- Helicopters need to make multiple stops.
- Altitude restrictions increase route distance.

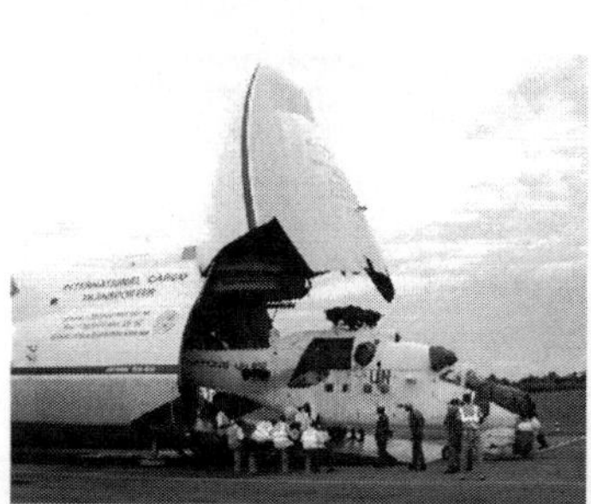

An An-124 unloads an Mi-24 (United Nations Movement Control photo, CC BY-SA 2.0)

Platform	Range (km)	Speed (km/hr)	Flight and refueling time (hrs)
Mi-8	983	224	18.5
Mi-24	1,000	269	19.5

NOTE: Availability of aircraft for personnel transport is not a stressing factor due to availability of nonmilitary assets. Therefore, this figure focuses on equipment.

Figure 2.5
Tajikistan Scenario Rail Assumptions

Assumption set 1 (rail asset demand)

MR BTGs, artillery and missile units, and MTO battalions move all equipment and class supplies by rail. Railcars and trains are readily available and in position when units are ready to deploy.

Assets to transport	Origin	Number of railcars (equipment + personnel)*	Number of trains
74th Guards MR Brigade, 120th Artillery Brigade	Yurga (Kemerovo Oblast)	297 (261 + 36)	5
35th MR Brigade	Aleysk	219 (189 + 30)	4
21st MR Brigade	Totskoye	227 (197 + 30)	4
106th MTO Brigade	Yurga (Kemerovo Oblast)	450 (420 + 30)	8
119th Missile Brigade	Elanskiy	53 (38 + 15)	1
Border troops	Central Military District	286 (226 + 60)	5
	Total	**1,532**	**27**

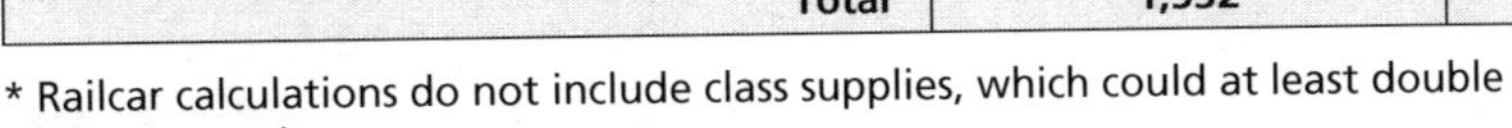

* Railcar calculations do not include class supplies, which could at least double railcar demand.

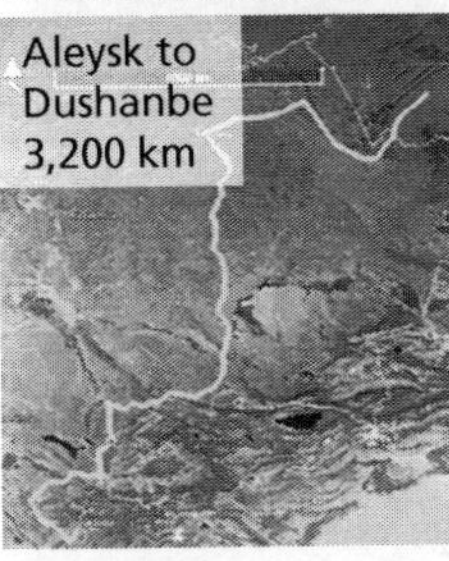

Assumption set 2 (rail closure)

- Rail line to Queb determined too high-risk due to proximity to Afghan border; all equipment is thus sent to Dushanbe.
- No routing issues due to bridge or tunnel limitations.
- Customs and clearances expedited at Kazakh, Uzbek, and Tajik borders.
- One day at origin rail loading point and travel speed of 40 km/hr.
- With 24-hour operations at Dushanbe, train load clears every 4 hours.

Closure of equipment and personnel takes at least 9 days

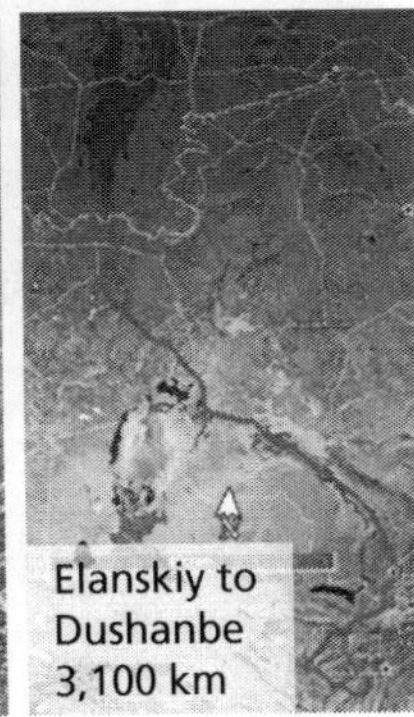

Figure 2.6
Tajikistan Scenario Movements to Tactical Assembly Areas

Assumptions
- Unopposed, faster administrative ground movement to TAAs
- Preestablished forward logistics areas
- Weight of effort distributed across all five TAAs

TAA	1	2	3	4	5
Distance (km)	455	281	195	161	128
Vehicles	658	658	658	658	658
Completion time (hrs)	20.0	11.0	8.0	6.8	5.6
Assumptions	• 50 m vehicle spacing during the day, 75 m at night • 30% of time spent stopped for rest/ maintenance • Road conditions will limit movement to an average of 40 km/h during the day, 30 km/hr at night				

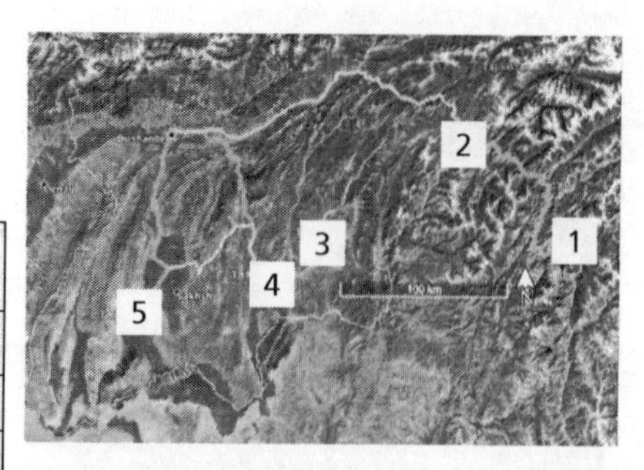

Photo by Alj87 via Wikimedia Commons (CC BY-SA 3.0)

Additional considerations
- Column length during the day will be ~40 km, ~60 km at night.
- Many of the roads to TAAs 3–5 are secondary roads, and the M41 highway has difficult terrain, slowing speed.
- Delays may be caused by harassment, attacks, weather, terrain, movement at night, or vehicle breakdowns.

NOTE: Estimates assume formation along a single road. Other options are stagger or diamond formations if road width allows.

Key Points from the Tajikistan Scenario

In this scenario, Russia benefits from its large existing base in Tajikistan and from its long-standing familiarity with the terrain and supply routes. Sustainment would be relatively easy, given existing facilities and storage. However, movement to the tactical assembly areas and areas of operation would be far more challenging than the initial waves of transportation. These movements would require navigating rough terrain, narrow passes, and long distances. Our scenario requires Russia to establish a second sustainment base in Kazakhstan to support operations in Tajikistan and Afghanistan.

CHAPTER THREE

Serbia Scenario

This is a small-footprint special operations deployment to respond to a notional attempt to overthrow the Serbian government. Russia deploys a small joint task force to an assembly area in Niš, Serbia, to enable follow-on movement and help defend government facilities and control violent protests in Belgrade and Novi Sad. After a covert insertion of the initial wave of forces is uncovered, Russia must deploy southwest of Serbia through a narrow geographic corridor at Neum in Bosnia and Herzegovina to bypass a NATO air blockade.

Figures 3.1–3.6 show, respectively, the deployment range, available Russian forces, movement plan, airlift assumptions, sealift assumptions, and ground movement assumptions.

Figure 3.1
Serbia Scenario Deployment Range

Figure 3.2
Russian Forces in the Serbia Scenario

GROUND FORCES

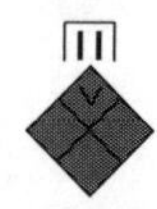

Joint Task Force Command
76th Air Assault Division
Headquarters Element

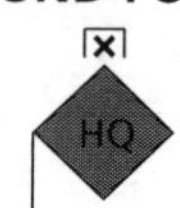

VDV (Airborne)

- 4,100 contract personnel
- 36 BMD-4M IFVs
- 20 BTR-MDM APCs
- 9 2S9 Nona mortars
- 12 D-30 howitzers
- 9 BTR-ZD APCs
- 6 BMD-1KSh IFVs
- 8 1V119 Reostat command vehicles
- 2 R-149 command vehicles
- 2 R-440 communications vehicles
- 20 support vehicles (heavy reliance on host-nation support)

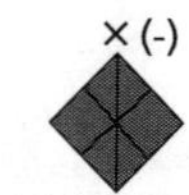

MR Brigade
(Collective Treaty Security Organization)

- 1,800 contract personnel
- 60 BTR-82AM APCs
- 6 BTR-80 APCs
- 15 MT-LBs* armored vehicles
- 12 2B9 Vasilek gun mortars
- 4 BRDM-2 patrol vehicles
- 6x ZSU 2S6M Tunguskas*
- 30 support vehicles (heavy reliance on host-nation support)

* Tracked and would require transport augmentation

SUPPORT

- MTO battalion (-)
- 800 personnel
- 300 vehicles
- Engineer company
- Electronic warfare detachment

Spetsnaz

- 1,200 contract personnel
- 25 BTR-80 APCs
- 12 Tigr/Lynx (light jeeps)
- 10 Ural Typhoon-Us
- 5 support vehicles (heavy reliance on host-nation support)

Total personnel	**7,900**
Combat vehicles	242
Support vehicles	355
Helicopters	0

Justification: Deploy a self-sustaining joint combat team capable of semi-independent operations in an allied country against violent protestors. Send 76th VDV Division personnel but one BTG of associated equipment.

Figure 3.3
Serbia Scenario Movement Plan

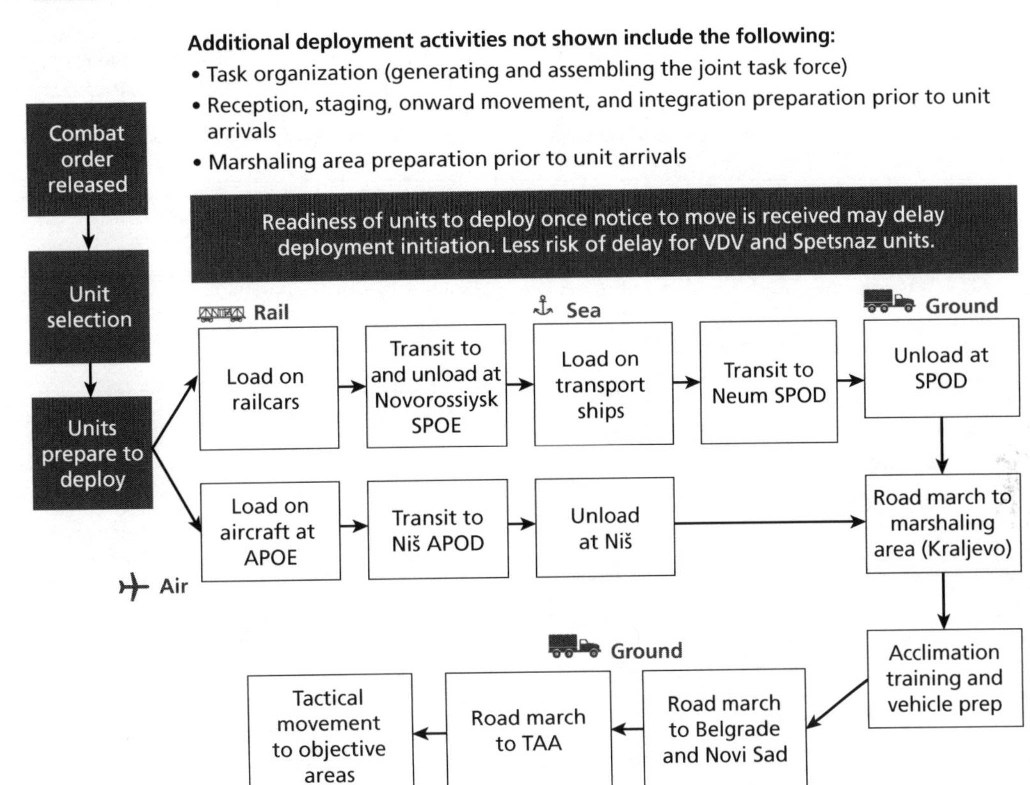

Figure 3.4
Serbia Scenario Airlift Assumptions

Assumption set 1 (affecting air asset demand in first wave: Spetsnaz)

- Russia sends equipment by military aircraft marked for Syrian humanitarian aid to Niš.
- Russia selects flight legs that allow for maximum cargo capacity.

Il-76

Number of sorties (equipment)	% of available fleet*
14	23

or

An-124

Number of sorties (equipment)	% of available fleet*
6	100

Sortie calculations are based on weight and do not include class supplies. Class supplies and loading factors will increase the number of sorties required.
* Assumes 60 of 110 Il-76s or 6 of 9 An-124s are available.

Platform	Number of sorties (personnel)	% of available fleet
Large aircraft	9	Not relevant**
Small aircraft	24	

**Russia has used civilian and other government aircraft to transport troops to Syria, in addition to its own military assets. Availability of transport for personnel is not as limiting a factor as it is for heavy lift assets

Assumption set 2 (affecting closure times in first wave: Spetsnaz):

- Because of covert insertion, denial of NATO overflight is not an issue.
- Flight and closure times are not stressing factors.
- Maximum 4 aircraft on the ground, 24-hour operations.
- Only Il-76s and heavy aircraft are used.

Romania
Bucharest
Serbia
Sofia
Bulgaria
Kosovo
Macedonia (FYROM)
Chisinau
Odessa
Krasnodar
1 leg; 1,395 km; 2-hour flight

Assumption set 3 (affecting air asset demand in second wave: VDV)

Russia selects flight legs that allow for maximum cargo capacity

Il-76

Number of sorties (equipment)	% of available fleet*
26	23

or

An-124

Number of sorties (equipment)	% of available fleet*
11	100

Sortie calculations are based on weight and do not include class supplies. Class supplies and loading factors will increase the number of sorties required.
* Assumes 60 of 110 Il-76s or 6 of 9 An-124s are available.

Platform	Number of sorties (personnel)	% of available fleet
Large aircraft	30	Not relevant, though small aircraft demand may stress available assets**
Small aircraft	83	

**Russia has used civilian and other government aircraft to transport troops to Syria, in addition to its own military assets. Availability of transport for personnel is not as limiting a factor as it is for heavy lift assets

Assumption set 4 (two cases):

1. Turkey allows overflight despite NATO refusal.
2. All NATO overflight is denied (Iraq allows).

Initial load at APOE takes 1 day. Assume all fly the same route at the same time (for simplicity). Malta allows stopping and overflight.

Mix of 6 An-124 and 12 Il-76s for equipment, large aircraft for personnel. Each airfield has a maximum of 4 aircraft on the ground and 24-hour operations (~27-hour clearance at each leg for refueling).

Figure 3.5
Serbia Scenario Sealift Assumptions

Assumption set 1 (affecting sea asset demand in third wave: MR BTGs)

- MR BTGs and MTO (-) send equipment and class supplies by sea and personnel by air.
- Without access to NATO seaport, Russia must use Bosnia-Herzegovina's ocean access at Neum, which does not have a sufficient port, cargo handling, or capacity for larger commercial vessels. Russia must therefore conduct beach landings with organic assets, complicated by steep terrain.
- Because of limited inventory, each available *Ropucha* and *Tapir/Alligator* must make multiple round-trips to close the force.

Ropucha-class landing ship
(U.S. European Command photo)

Options to move 450 vehicles and initial class supplies

Project 1171 (*Tapir/Alligator*)

Number of sorties	% of available fleet*
~10	333

or

Project 775 (*Ropucha*)

Number of sorties	% of available fleet*
~17	243

* Assumes 7 of 15 *Ropuchas* and 3 of 4 *Tapir/Alligators* are available.

Assumption set 2 (affecting sea closure time in third wave: MR BTGs)

- Turkey allows passage through Bosporus Strait, allowing use of SPOE at Novorossiysk.
- 24-hour load time at SPOE, 36-hour unload time at SPOD, and travel at 18 knots.
- Maximum of 3 vessels can load and unload at a time.
- Russia uses all 3 available *Tapir/Alligators* (2 round trips each), 3 *Ropuchas* (2 round trips each), and remaining 4 *Ropuchas* (1 trip each).

Figure 3.6
Serbia Scenario Ground Movement Assumptions

Assumption set 1 (administrative ground movement)

50 m vehicle spacing, 50 km/hr, and 20% time spent halted for rest/maintenance/security.

Assumption set 2

VDV forces move before MR BTGs

Potential delays to optimal unit travel times may be caused by harassment or attacks, weather, terrain, movement at night as opposed to day, or vehicle breakdowns.

The latter is more likely if units' deployment to this stage sacrificed post-sealift vehicle maintenance to expedite onward movement.

Convoy route	Niš to Novi Sad (417 km)	Niš to Belgrade (239 km)	Neum to Novi Sad (482 km)	Neum to Belgrade (484 km)
Wave	VDV	VDV	MR BTGs	MR BTGs
Vehicles*	62	62	225	225
Column length (km)**	3.72	3.72	13.50	13.50
Completion time (hours)	10.5	6.0	13.9	14.0

*Assumes effort split evenly between both. As shown in the maps, columns try to avoid similar routes to reduce road congestion

**Assumes file formation along single road. Multiple roads may be taken, but they must be secured. Coordination of convoy would also be more difficult.

Key Points from the Serbia Scenario

This scenario highlights the limits imposed by international restrictions. In this case, Serbia is a short geographic distance from Russia's Western Military District, but it is effectively nested among NATO countries. Using the narrow pathway from Neum would be practical only for a small force, not for a major deployment. Russia could try to bully its way into Serbia, but it would risk triggering a NATO Article 5 contingency. Absent sufficient access, even such a near deployment becomes quite challenging for Russian ground forces.

CHAPTER FOUR

Syria Scenario

When we developed this scenario in early 2017, Russia was continuing to support the Syrian armed forces' operations against various insurgent and terrorist groups. For the notional 2023 scenario, we selected an internal deployment location within Syria (Palmyra) that was far enough away from the main Russian bases in the northwest of the country to stress Russian capabilities. Notionally, a Russian Spetsnaz unit is encircled by a large, well-armed insurgent force within and around Palmyra. Syrian ground combat units supporting the Spetsnaz unit are incapable of breaking through to rescue or reinforce the trapped Russian soldiers. Russian ground forces in Syria are otherwise engaged in vital security missions, so Russia deploys a brigade combat team to its airfield at Khmeimim, its seaport at Latakia, and then over ground to Palmyra.

This is one of two *far* scenarios. Figures 4.1–4.6 show, respectively, the deployment range, available Russian forces, movement plan, airlift assumptions, sealift assumptions, and ground movement assumptions.

Figure 4.1
Syria Scenario Deployment Range

Figure 4.2
Russian Forces in the Syria Scenario

GROUND FORCES

Joint Task Force Command
27th MR Brigade and brigade headquarters element

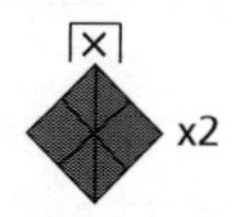

2 MR BTGs, each with

- 50 BTR-82/82A APCs
- 8 120-mm mortars
- 10 T-90A main battle tanks
- 2 2S6M1 (SA-19) air defense vehicles
- Organic sustainment

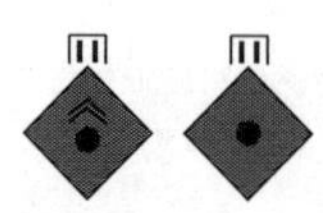

Brigade artillery group with

- Artillery reconnaissance element
- Cannon battalion (attached) with 18 towed 2A65 152-mm howitzers
- MRL battalion with 18 Tornado-Gs
- 2 2S6M1 (SA-19) air defense vehicles
- Organic sustainment

AIR

Joint Air Wing

- Rotary-wing attack aviation in country
- Unmanned systems: Forpost and Orlan-10

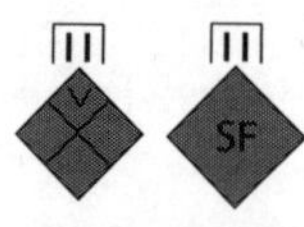

Airborne Forces

Air assault battalion

- 32 BMD-4M IFVs
- 2 BTR-MDM APCs
- Organic sustainment

Spetsnaz element

- 3 companies
- 1 communications company
- 7 BTR-80 APCs
- 3 Tigrs (light jeeps)
- 3 Tayfun-Us (MRAPs)

SUPPORT

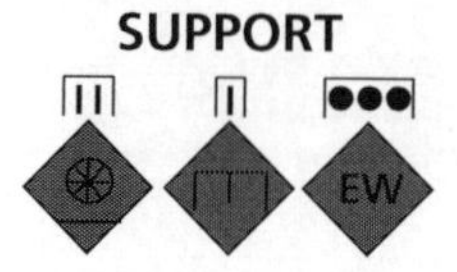

Other combat support and sustainment elements

- Engineer company
- MTO battalion
  - 408 vehicles
  - 1,190 tons of dry supplies
  - 680 tons of liquids
  - Additional food, fuel, and ammunition
- Electronic warfare detachment

Total personnel	4,666
Combat vehicles	211
Support vehicles	~500
Helicopters	20

NOTE: MRAP = mine-resistant, ambush-protected (vehicle).

Figure 4.3
Syria Scenario Movement Plan

General steps of the joint task force's movement by air and sea to Syria, as well as en route to the objective area at Palmyra
Two waves of deployment: VDV and Spetsnaz units completely by air, followed by major combat forces arriving by both sea (equipment) and air (personnel).

Additional deployment activities not shown include the following:
- Task organization (generating and assembling the joint task force)
- Reception, staging, onward movement, and integration preparation prior to unit arrivals
- Marshaling area preparation prior to unit arrivals

Combat order released → Unit selection → Units prepare to deploy

Readiness of units to deploy once notice to move is received may delay deployment initiation. Less risk of delay for VDV and Spetsnaz units.

Sea — Heavy combat and support equipment, some class supplies:
Load on railcars → Transit to and unload at SPOE → Load on transport ships → Transit to Syrian SPOD(s) → Unload at SPOD(s)

Air — Personnel, airborne gear, some class supplies:
Load on aircraft at APOE → Transit to Syrian APOD(s) → Unload at APOD(s) → Road march to marshaling area

Ground:
Acclimation training and vehicle prep → Road march to Tiyas → Road march to TAA → Tactical movement to objective area

Figure 4.4
Syria Scenario Airlift Assumptions

Assumption set 1 (initial ground force deployment)

- VDV and Spetsnaz move personnel, equipment, and some class supplies by air.
- Loading aircraft at APOE takes 1 day.
- Airlift for VDV, Spetsnaz, and aviation assets and personnel uses Khmeimim Air Base only as APOD, with estimated maximum 4 aircraft on the ground at a time and 24-hour operations.
- Turkey allows overflight.

Closure time of initial wave (VDV and Spetsnaz equipment and personnel): 3–4 days after unit is ready to deploy

Assets to Lift	Equipment*	Personnel
VDV sorties	12 Il-76s or 5 An-124s	3 large or 7 small transport aircraft
Spetsnaz sorties	5 Il-76s or 2 An-124s	2 large or 5 small transport aircraft
Sortie totals	17 Il-76s or 7 An-124s	5 large or 11 small troop transport equivalents
% of available fleet**	28% of Il-76s 117% of An-124s	Not relevant: Availability for personnel not a limiting factor***

* Sortie calculations are based on weight and do not include class supplies. Class supplies and loading factors increase the number of sorties required.

** Assumes 60 of 110 Il-76s or 6 of 9 An-124s are available.

*** Russia has used civilian and other government aircraft to transport troops to Syria, in addition to its own military assets. Availability of transport for personnel is not as limiting a factor as it is for heavy lift assets

Assets to Lift	Personnel
27th MR Brigade (-) sorties	24 large troop transport aircraft (Il-76 equivalent) or 63 small (An-24 equivalent) troop transport aircraft
% of available fleet	Not relevant: Availability for personnel not a limiting factor

Assumption set 2 (follow-on ground force deployment)

- Follow-on units' equipment and some class supplies move by sea. All personnel and some class supplies move by air.
- Loading aircraft at APOE takes 1 day.
- Airlift for 27th MR Brigade (-) and support unit personnel uses Latakia Air Base only as APOD, with estimated maximum 4 aircraft on the ground at a time.
- Turkey allows overflight.

Closure time of initial wave (VDV and Spetsnaz equipment and personnel): 3–4 days after unit is ready to deploy

Figure 4.5
Syria Scenario Sealift Assumptions

Assumption set 1 (sea asset demand in second wave)
Second-wave forces—27th MR Brigade (-) and support units—transport all equipment and some class supplies by sea. Northern and Baltic fleets have diverted transport assets to assist. Deployment includes both organic and nonmilitary cargo vessels, particularly roll-on/roll-off ships.

Total Cargo	808 vehicles, class supplies
SPOE	Novorosleseksport
Sorties*	~30 *Tapirs*/*Alligators* or ~60 *Ropuchas* or 3–4 nonmilitary vessels
% of available assets*	1,000% of *Tapirs*/*Alligators*, 857% of *Ropuchas*; less stressing for nonmilitary assets**

* Assumes military vessels are not used for class supplies; using nonmilitary vessels for this purpose in parallel would not be a time stress factor.

**Assumes that 3 or 4 total *Tapirs*/*Alligators* and 7 of 15 *Ropuchas* are available.

Assumption set 2 (closure times of second wave)
- Rail unloading at SPOE takes 1 day for initial arrival.
- Military vessels can travel at 18 knots; nonmilitary at 10 knots.
- Both Latakia and Tartus SPODs are used to alleviate backup. All deployment ports can accommodate 4 medium **or** 2 large roll-on/roll-off vessels at a time.

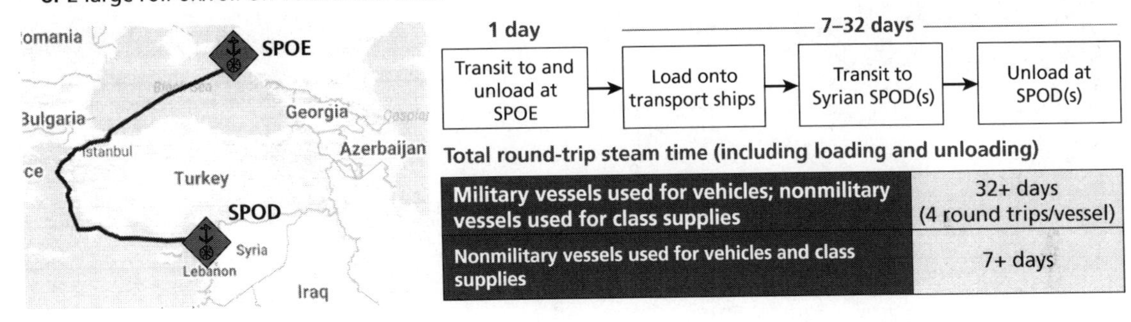

Total round-trip steam time (including loading and unloading)

Military vessels used for vehicles; nonmilitary vessels used for class supplies	32+ days (4 round trips/vessel)
Nonmilitary vessels used for vehicles and class supplies	7+ days

Figure 4.6
Syria Scenario Ground Movement Assumptions

Administrative Movement to Tiyas TAA

Convoy Route	Khmeimim to Tiyas (260 km)	Tartus to Tiyas (195 km) OR	Latakia to Tiyas (277 km)
Wave	VDV+ Spetsnaz	27th MR Brigade (-) + support	
Vehicles	68	808	
Column length* (km)	4.0	48.5	
Completion time (hours)	6.6	5.9	7.9
Assumptions	50 m vehicle spacing at 50 km/h; 20% time spent stopped for rest/maintenance		

Assumption set 1
- Unopposed, faster administrative ground movement to Tiyas TAA.
- No anticipated terrain or weather delays for ground or air units.
- Requires preestablished forward logistics base at Tiyas Airfield (T-4).

Assumption set 2
VDV/Spetsnaz move prior to second-wave force to establish reconnaissance and other preparatory activities in Tiyas.

Tactical Movement to Tiyas TAA

Convoy Route	VDV + Spetsnaz	27th MR Brigade (-) + support
Vehicles	68	808
Column length* (km)	6.1	72.7
Completion time (hours)	2.8	5.0
Assumptions	75 m vehicle spacing at 30 km/hr; 20% time spent stopped for rest/maintenance/security	

* Assumes formation along single road. Parallel road to Tiyas would cut time, decrease column length. Other options are stagger and diamond techniques if road width allows.

Assumption set 1
- Deliberate, slower tactical ground movement from Tiyas to Palmyra.
- No anticipated terrain or weather delays for ground or air units.

Assumption set 2
VDV/Spetsnaz move prior to second-wave force to establish reconnaissance and other preparatory activities in Palmyra.

Delays may be caused by harassment, attacks, weather, terrain, movement at night, or vehicle breakdowns (more likely if post-sealift vehicle maintenance is sacrificed to expedite onward movement).

Key Points from the Syria Scenario

Even with a fairly robust basing system and existing forces in theater, deployment from Russia into theater and then into combat proves challenging. Airlifting forces into Russian bases on Syrian soil is fairly easy, if time-consuming. Dropping those forces near the objective area from aircraft would be a viable alternative, but airdropped forces would have less available combat power than forces deployed by sea and air into ports of debarkation, assembled, and moved forward over ground.

CHAPTER FIVE

Venezuela Scenario

In this *far* scenario, the Venezuelan government requests Russian assistance in putting down increasingly violent protests in Caracas, and Venezuela is on the verge of collapse. Russia deploys a joint task force of 7,000 personnel in the form of a motorized infantry brigade and a light naval squadron. The primary threats to the task force are armed gangs and large civilian protests that might include armed instigators. Figures 5.1–5.6 show, respectively, the deployment distance, available Russian forces, movement plan, airlift assumptions, sealift assumptions, and ground movement assumptions.

Figure 5.1
Venezuela Scenario Deployment Range

Figure 5.2
Russian Forces in the Venezuela Scenario

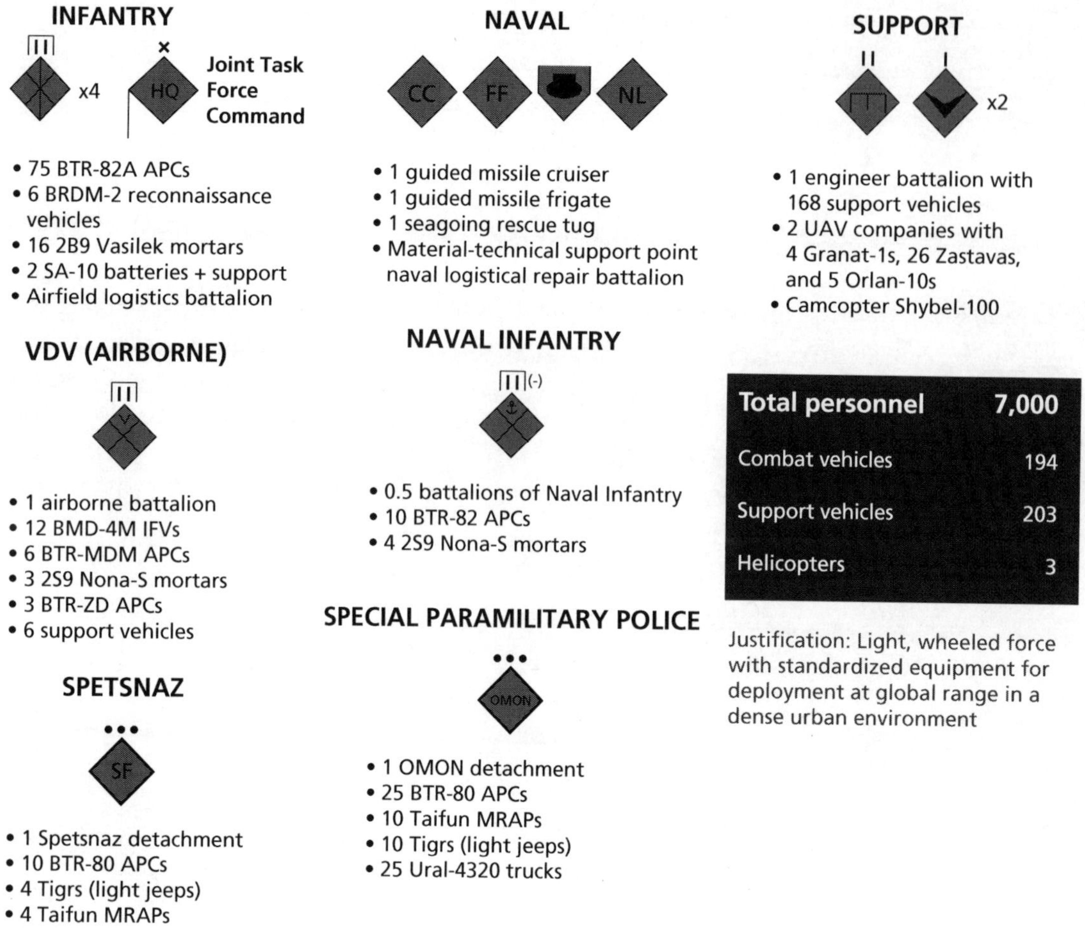

INFANTRY

x4 Joint Task Force Command

- 75 BTR-82A APCs
- 6 BRDM-2 reconnaissance vehicles
- 16 2B9 Vasilek mortars
- 2 SA-10 batteries + support
- Airfield logistics battalion

VDV (AIRBORNE)

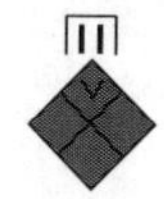

- 1 airborne battalion
- 12 BMD-4M IFVs
- 6 BTR-MDM APCs
- 3 2S9 Nona-S mortars
- 3 BTR-ZD APCs
- 6 support vehicles

SPETSNAZ

- 1 Spetsnaz detachment
- 10 BTR-80 APCs
- 4 Tigrs (light jeeps)
- 4 Taifun MRAPs

NAVAL

- 1 guided missile cruiser
- 1 guided missile frigate
- 1 seagoing rescue tug
- Material-technical support point naval logistical repair battalion

NAVAL INFANTRY

- 0.5 battalions of Naval Infantry
- 10 BTR-82 APCs
- 4 2S9 Nona-S mortars

SPECIAL PARAMILITARY POLICE

- 1 OMON detachment
- 25 BTR-80 APCs
- 10 Taifun MRAPs
- 10 Tigrs (light jeeps)
- 25 Ural-4320 trucks

SUPPORT

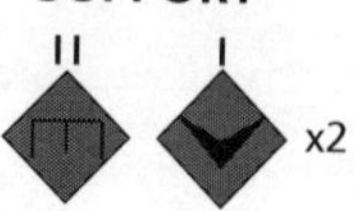

- 1 engineer battalion with 168 support vehicles
- 2 UAV companies with 4 Granat-1s, 26 Zastavas, and 5 Orlan-10s
- Camcopter Shybel-100

Total personnel	7,000
Combat vehicles	194
Support vehicles	203
Helicopters	3

Justification: Light, wheeled force with standardized equipment for deployment at global range in a dense urban environment

Figure 5.3
Venezuela Scenario Movement Plan

General steps of the joint task force's movement by air and sea to Venezuela, as well as en route to the objective area at Caracas

Deployment occurs in two waves: VDV and Spetsnaz units completely by air, followed by major combat forces arriving by both sea (equipment) and air (personnel)

Additional deployment activities not shown include the following:

- Task organization (generating and assembling the joint task force)
- Reception, staging, onward movement, and integration preparation prior to unit arrivals
- Marshaling area preparation prior to unit arrivals

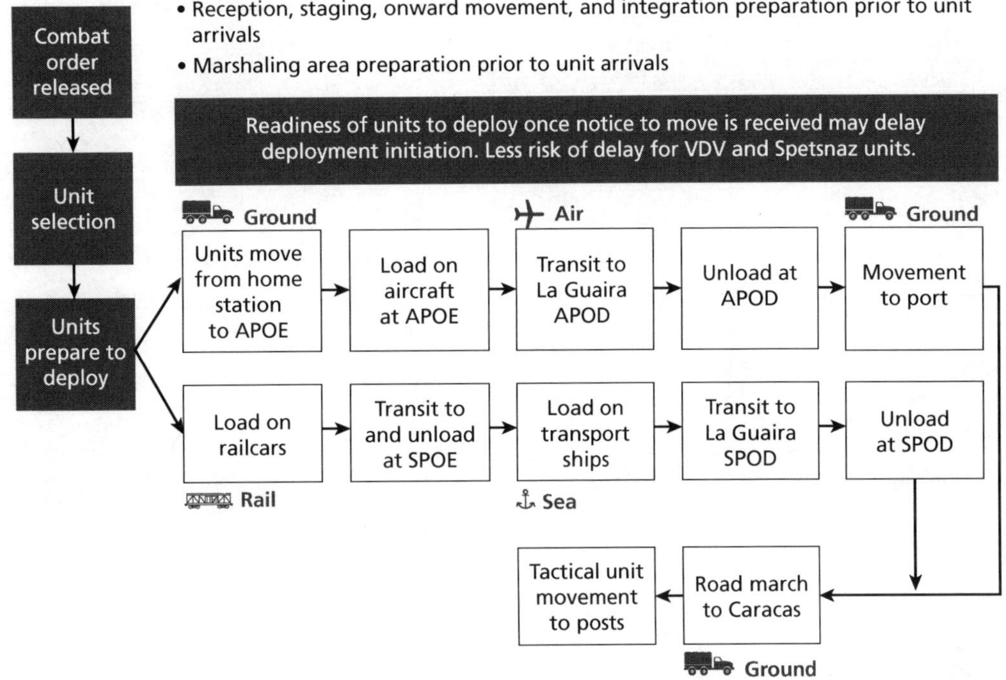

Figure 5.4
Venezuela Scenario Airlift Assumptions

Assumption set 1 (air asset demand in initial wave)
Airborne BTG, Spetsnaz, and Naval Infantry move personnel, equipment, and some class supplies by air. The major impact on air asset demand is the leg from Casablanca to Venezuela; long range reduces the amount of cargo that the aircraft can carry.

	Platform	Max. cargo at 6,600 km (metric tons)	Airborne BTG sorties*	Special operations forces sorties*	Naval Infantry sorties*	% of available fleet
Equipment	Il-76	26	16.3	10.5	7.2	57
	An-124	95	4.4	2.9	2.0	155
	Platform		Sortie requirements for personnel			
Personnel	Large aircraft		4.6	2.9	1.8	Not relevant: Availability for personnel not a limiting factor
	Small aircraft		13.0	8.0	5.0	

*Sortie calculations are based on weight and do not include class supplies. Class supplies and loading factors increase the number of sorties required.

Assumption set 2 (affecting closure times of initial wave)

- Overflight and basing access are critical because in-air refueling capabilities are inadequate.
- Initial load at APOE takes 1 day.
- All fly the same route at the same time (for simplicity).
- Each airfield in leg has maximum 4 aircraft on the ground at a time (~15-hour clearance at each leg for refueling if using 6 An-124s and 12 Il-76s for equipment, large aircraft for personnel).

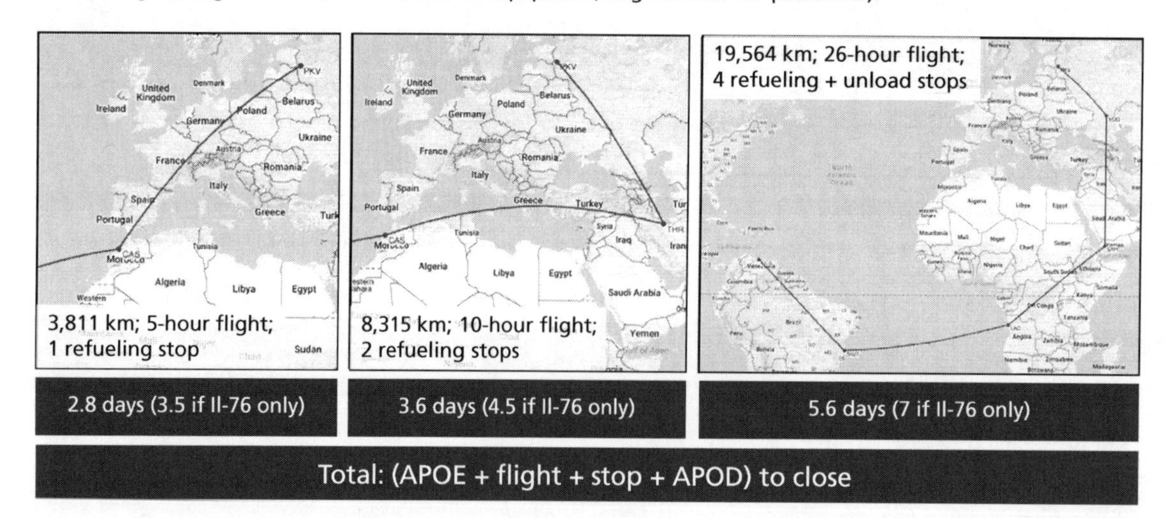

Figure 5.5
Venezuela Scenario Sealift Assumptions

Assumption set 1 (affecting sea asset demand in second wave)

- MR BTGs and OMON units send all equipment and class supplies by sea.
- Personnel travel by air using nonmilitary transport.
- Organic support vessels (*Tapir/Alligator*, *Ropucha*) cannot be used over such long distances. Nonmilitary or commercial may be required; roll-on/roll-off vessels bought for Syria Express (*Alexandr Tkachenko* and suspected MV *Novorossiysk*) may be used, but enduring high rates of usage may have degraded their readiness.
- Limiting factor is not space but time to acquire nonmilitary vessels and steam time. Hiring commercial vessels may take several days to weeks, depending on the company.

Assumption set 2 (affecting sea asset closure rates in second wave)

- Novorossiysk and Murmansk ports are preferred as SPOEs. Murmansk determined to be less politically risky in case of denied access to Bosporus Strait.
- Loading at SPOE takes 1 day, unloading at SPOD takes 2 days each.

Total round-trip steam time (including loading and unloading)

Military vessels used for vehicles; nonmilitary vessels used for class supplies	NA
Nonmilitary vessels used for vehicles and class supplies	16–18 days

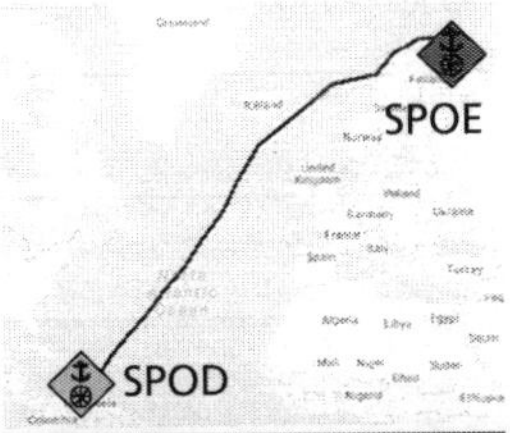

SPOE
SPOD
Murmansk
9,870 km; 16 days

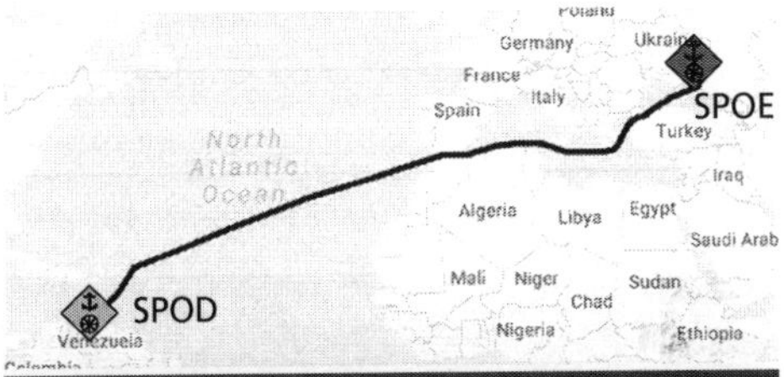

SPOE
SPOD
North Atlantic Ocean
Germany
France
Spain
Italy
Turkey
Iraq
Algeria
Libya
Egypt
Saudi Arab
Mali
Niger
Chad
Sudan
Nigeria
Ethiopia
Venezuela
Novorossiysk
11,230 km; 18 days

Figure 5.6
Venezuela Scenario Ground Movement Assumptions

Tactical Movement to Forward Operating Bases

Convoy Route	Area to Forward Operating Base Fort Tiuna (25 km)	Maiquetia Marshaling Area to Francisco de Miranda Air Base (24 km)	Maiquetia Marshaling Area to Forward Operating Base Fort Tiuna (24 km)	Maiquetia Marshaling Area to Francisco de Miranda Air Base (24 km)
Wave	Airborne, special operations forces, Naval Infantry	Airborne, special operations forces, Naval Infantry	MR, special paramilitary police	MR, special paramilitary police
Vehicles*	38	39	94	95
Column length (km)	3.4	3.5	8.5	8.5
Completion time (hours)	1.2	1.1	1.3	1.3

*Assumes weight of effort split evenly between both.

Assumption set 1

- Deliberate, slower tactical ground movement
- 75m vehicle spacing, traveling at 30km/hr, 20% of time spent halted for rest/maintenance/security measures

Assumption set 2

VDV/Spetsnaz move prior to second-wave force to establish reconnaissance and other preparatory activities in forward operating bases.

Potential delays to optimal unit travel times may be caused by harassment or attacks, weather, terrain, movement at night as opposed to day, or vehicle breakdowns. The latter is more likely if units' deployment to this stage sacrificed post-sealift maintenance on vehicles to expedite onward movement.

Key Points from the Venezuela Scenario

This is the longest notional scenario that we considered. Air movement requires two interim stops for each sortie, while sealift would require at least 16 days of sailing time. Both these movements are highly dependent on in-transit movement authorities and refueling options and, therefore, diplomatic largesse. The absence of a network of alliances and international bases significantly increases the likelihood that this deployment would suffer setbacks or delays. And the lack of long-range sustainment would greatly complicate Russia's ability to keep its ground forces fueled, fed, watered, and sufficiently supplied with ammunition over time, particularly as Caracas suffers from acute shortages of various classes of supplies.

CHAPTER SIX

Ukraine "+1" Scenario

We developed and studied—but did not analyze—this notional scenario as part of our collective assessment. This is a large-scale Russian military invasion of Ukraine involving approximately 130,000 Russian joint force personnel, centering on an RGF task force of approximately 83,000 soldiers built around the 20th Combined Arms Army and the 8th Combined Arms Army.

Figures 6.1–6.6 show, respectively, the deployment distance, available Russian forces, operational phase 1, operational phase 2, timeline of units ready to deploy, and rail assumptions.

Figure 6.1
Ukraine Scenario Deployment Range

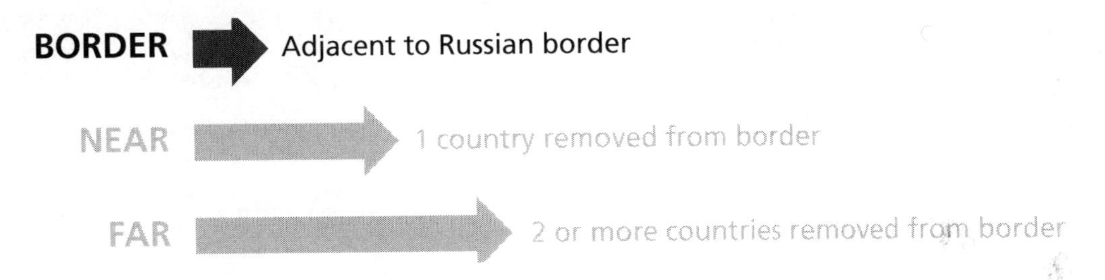

Figure 6.2
Russian Forces in the Ukraine Scenario

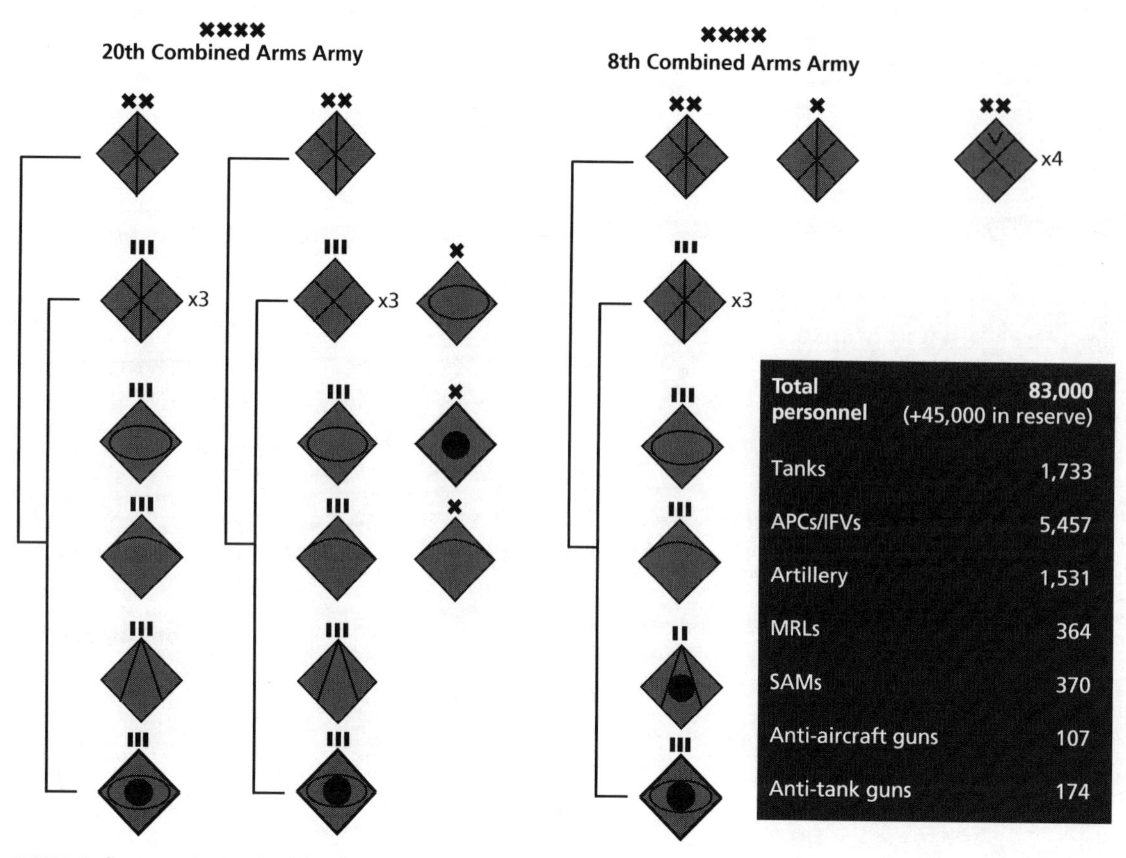

NOTE: Reflects authorized table of organization and equipment for units; Russia may not deploy all equipment.

Figure 6.3
Ukraine Scenario, Operational Phase 1

- Western Military District units assemble near Belgorod.
- Southern Military District units assemble near Taganrog and Novocherkassk.
- Initial special operations and Spetsnaz units inserted.

Russian units operate in their overt and traditional capacity, with their traditional logistics requirements.

- Will be too large and complex to be a covert "hybrid" conflict
- Some hybrid activity anticipated

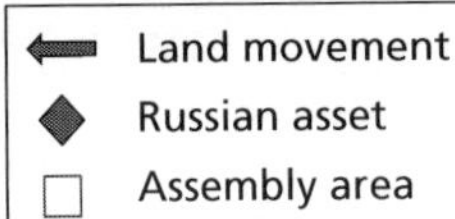

Figure 6.4
Ukraine Scenario, Operational Phase 2

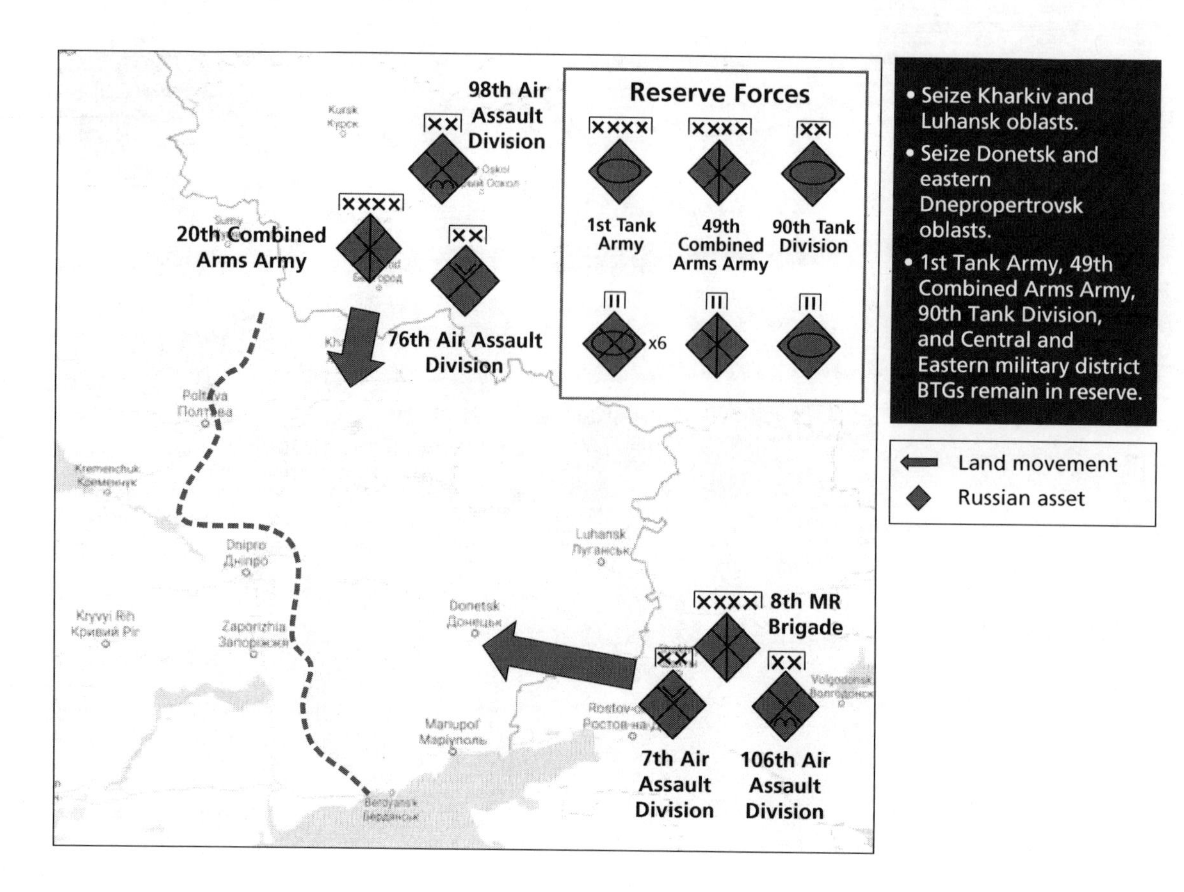

Figure 6.5
Timeline of Units Ready to Deploy in the Ukraine Scenario

Units	Ready to deploy within 10 days	Ready to deploy in 10–20 days	Ready to deploy in 30 days
Western Military District	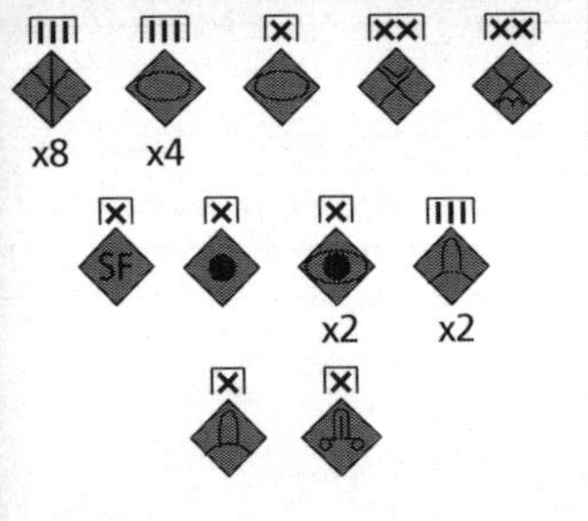 x8 x4 SF x2 x2	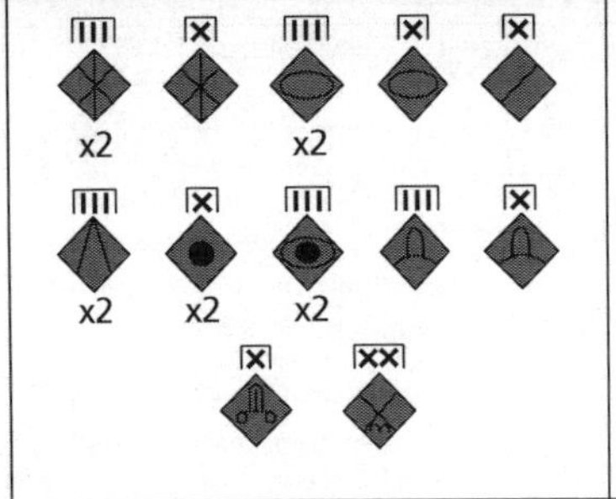 x2 x2 x2 x2 x2	
Southern Military District	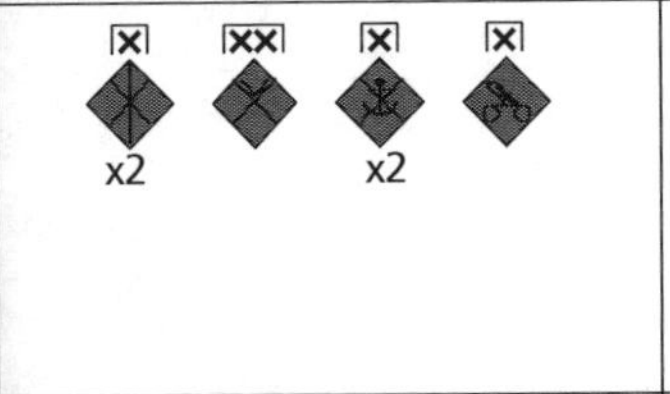 x2 x2	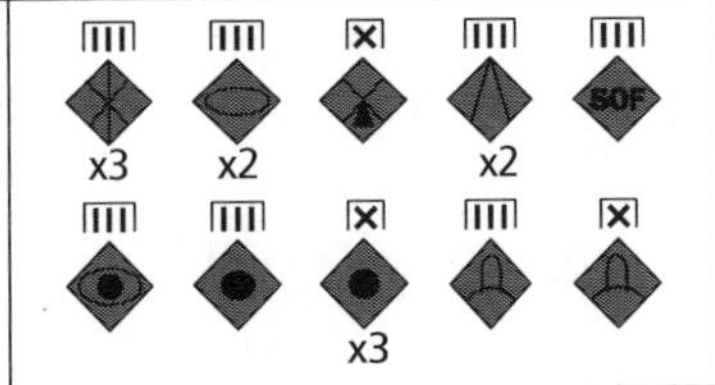 x3 x2 SOF x2 x3	
Central Military District	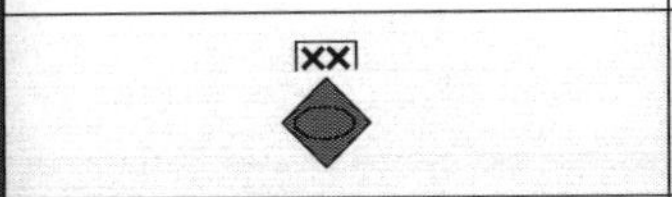		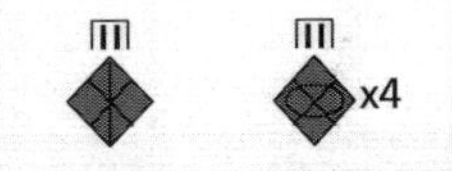 x4
Eastern Military District			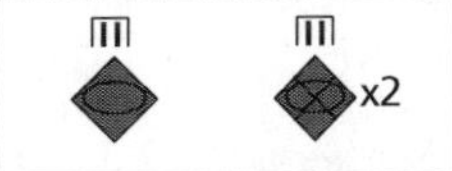 x2
Total combat vehicles	**4,155 tracked 1,600 wheeled**	**2,393 tracked 983 wheeled**	**506 tracked 243 wheeled**

- Operations do not have a rolling start. Russia builds forces along the border with Ukraine before entering.
  - Because Russia chooses the time/place of the operation, it can deploy half of its initial combat force from the Western and Southern military districts within 10 days of order and the other half within 20 days of order.
  - Phased deployment will allow local units to train for combat missions along the border with Ukraine before entering.
- Due to mixed manning in Russian units, most will not deploy in the same wave (e.g.,1–2 regiments in a division for wave 1, the remainder in wave 2), with a few exceptions for high-readiness units like the VDV, which deploy at once.

Figure 6.6
Ukraine Scenario Rail Assumptions

Assumption set 1 (rail asset demand)
The military is given priority order for railcars and trains per Resolution No. 761 of October 7, 1998.

Readiness Wave	Military District	Railcars/ Trains (all vehicles)	Railcars/Trains (tracked vehicles only)
Within 10 days	Western	2,411 / 43	1,963 / 35
	Southern	811 / 15	573 / 11
	Central	402 / 8	257 / 5
	TOTAL	**3,624 / 66**	**2,859 / 51**
Within 10–20 days	Western	1,388 / 25	1,003 / 18
	Southern	910 / 16	707 / 13
	TOTAL	**2,298 / 41**	**1,710 / 31**
Within 30 days	Central	292 / 6	203 / 4
	Southern	205 / 4	159 / 4
	TOTAL	**497 / 10**	**362 / 8**

Calculations consider combat vehicles only. Support vehicles (which often number at least 1:1 in force packages), class supplies, and personnel would double or triple these estimates. Many assets can and should be road-marched to alleviate congestion and railcar shortages, however.

- Western Military District has the densest rail network in the country, but only two routes lead to Belgorod assembly area, causing congestion.
- Southern Military District rail network is less dense, but train density will be lower than in Western Military District.
- Central and Eastern military districts' rail networks are least dense and travel times are much greater, but train density will be low.

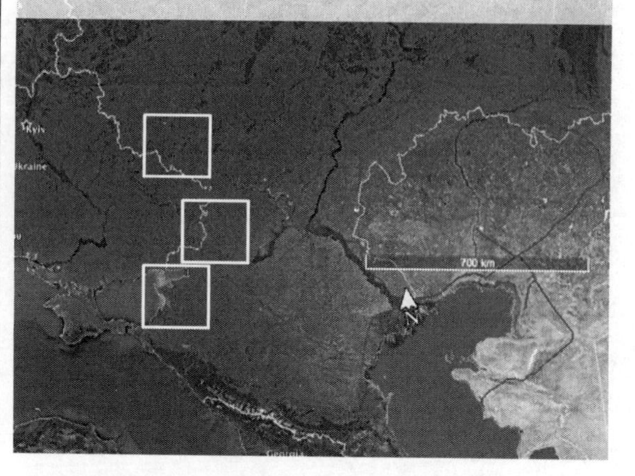

Assumption set 2 (rail closure)
- The military is given priority movement across Russia's rail network.
- Adequate crews and equipment are available for loading and unloading.
- Closure of first two waves will take **at least 28–30 days**

Key Points from the Ukraine Scenario

This is a large-scale operation that would take weeks, if not months, to fully develop and execute. There is little chance that Russia would be able to achieve operational surprise without undertaking significant hybrid warfare activities and warning observers. However, in many ways, this is a deployment sweet spot for the RGF: It is adjacent to Russia's Western and Southern military districts, where the core of its assets are located; support can travel across relatively flat terrain; it requires no transit across or around a hostile state; and it depends on relatively few joint transportation assets. If this is an ideal case, then the timeline should help inform future analyses of prospective Russian combat operations in Eastern Europe.

Bibliography

The following are key sources that informed our scenario development and analysis, as well as our review of 15 historical deployments of Soviet and Russian ground forces since 1945. See the companion report, *Russia's Limit of Advance: Analysis of Russian Ground Force Deployment Capabilities and Limitations* (available at www.rand.org/pubs/research_reports/RR2563), for detailed findings from our analysis of these notional and historical deployments.

Notional Scenarios

Background and Context for the Notional Scenarios

Al Jazeera, "ISIL: Target Russia," May 27, 2017. As of July 22, 2019:
http://www.aljazeera.com/programmes/specialseries/2017/05/isil-target-russia-170522095304580.html

Anceschi, Luca, and Bruce Pannier, "Is This Kazakhstan's New Transitional Government?" *Radio Free Europe*, September 23, 2016. As of July 22, 2019:
https://www.rferl.org/a/is-this-kazakhstans-new-transitional-government/28008984.html

Azami, Dawood, "Why Taliban Special Forces Are Fighting Islamic State," BBC News, December 18, 2015. As of July 22, 2019:
http://www.bbc.com/news/world-asia-35123748

Barabanov, Mikhail, ed., Новая армия России [*Russia's New Army*], Moscow: Centre for Analysis of Strategies and Technologies, 2010. As of July 22, 2019:
http://cast.ru/upload/iblock/df4/df4de3c02c0a4a3b7b6e6f9733c02382.pdf

Barabanov, M., and A. Vasiliev, Сирийский рубеж [*Syrian Frontier*], Moscow: Centre for Analysis of Strategies and Technologies, 2016. As of July 22, 2019:
http://cast.ru/upload/iblock/686/6864bf9d4485b9cd83cc3614575e646a.pdf

Baranets, Victor, "Генерал-полковник Андрей Картаполов: У России может появиться база в Сирии" ["Colonel-General Andrei Kartapolov: Russia May Have a Base in Syria"], *Komsomolskaya Pravda*, October 16, 2015. As of July 22, 2019:
https://www.kp.ru/daily/26446/3316981

Caravanistan, "Afghanistan Border Crossings," webpage, last updated July 3, 2019. As of July 22, 2019:
http://caravanistan.com/border-crossings/afghanistan

Central Intelligence Agency, "Kazakhstan," *The World Factbook*, last updated July 10, 2019. As of July 22, 2019:
https://www.cia.gov/library/publications/resources/the-world-factbook/geos/kz.html

Ch-Aviation, "Antonov to Sever Volga-Dnepr Ties by Early 2017," September 12, 2016. As of July 22, 2019:
https://www.ch-aviation.com/portal/news/49183-antonov-to-sever-volga-dnepr-ties-by-early-2017

"'Collective East': Why Russia Is Reinforcing Its Military Bases in Central Asia," *Sputnik*, November 6, 2017. As of July 22, 2019:
https://sputniknews.com/russia/201706111054527261-russia-bases-sco

"Военнослужащие по призыву не будут участвовать в боевых действиях" ["Conscripts Will Not Serve in Combat Roles"], RIA Novosti, February 14, 2013. As of July 22, 2019:
https://ria.ru/defense_safety/20130214/922957676.html

Defense Intelligence Agency, *Russia Military Power: Building a Military to Support Great Power Aspirations*, Washington, D.C., DIA-11-1704-161, 2017. As of July 22, 2019:
http://www.dia.mil/Portals/27/Documents/News/Military%20Power%20Publications/Russia%20Military%20Power%20Report%202017.pdf

Di Cocco, Alexandra, "Kazakhstan's Niche in China's Belt and Road Initiative," blog post, Atlantic Council, May 23, 2017. As of July 22, 2019:
http://www.atlanticcouncil.org/blogs/new-atlanticist/kazakhstan-s-niche-in-china-s-belt-and-road-initiative

Dizard, John, "Russians Jostle to Secure Money Lent to Venezuela," *Financial Times*, July 28, 2017. As of July 22, 2019:
https://www.ft.com/content/38103280-72ef-11e7-aca6-c6bd07df1a3c

Dolan, Daniel, "Opinion: Russian Tank Deal with Nicaragua 'Back to the Future' Moment for U.S.," U.S. Naval Institute, May 9, 2016. As of July 22, 2019:
https://news.usni.org/2016/05/09/opinion-russian-tank-deal-with-nicaragua-back-to-the-future-moment-for-u-s

Douglas, Nadja, "Civil-Military Relations in Russia: Conscript vs. Contract Army, or How Ideas Prevail Against Functional Demands," *Journal of Slavic Military Studies*, Vol. 24, No. 4, 2014, pp. 511–532.

Fiegel, Brenda, "Growing Military Relations Between Nicaragua and Russia," *Small Wars Journal*, December 5, 2014. As of July 22, 2019:
https://smallwarsjournal.com/jrnl/art/growing-military-relations-between-nicaragua-and-russia

Frolovskiy, Dmitry, "Kazakhstan's China Choice," *The Diplomat*, July 6, 2016. As of July 22, 2019:
http://thediplomat.com/2016/07/kazakhstans-china-choice

Grau, Lester W., and Charles K. Bartles, *The Russian Way of War: Force Structure, Tactics, and Modernization of the Russian Ground Forces*, Fort Leavenworth, Kan.: Foreign Military Studies Office, U.S. Army Combined Arms Center, 2016.

Haines, John R., "Everything Old Is New Again: Russia Returns to Nicaragua," Foreign Policy Research Institute, July 22, 2016. As of July 22, 2019:
https://www.fpri.org/article/2016/07/everything-old-new-russia-returns-nicaragua

International Institute for Strategic Studies, "Asia," *The Military Balance 2017*, London, January 2017, pp. 237–350.

———, "Russia and Eurasia," *The Military Balance 2017*, London, January 2017, pp. 183–236.

"ISIS in Afghanistan," video clip, *Frontline*, Season 33, Episode 8, November 17, 2015. As of July 22, 2019:
http://www.pbs.org/video/2365608927

Lavrov, Anton, "Towards a Professional Army," *Moscow Defense Brief*, No. 4 (48), April 2015. As of July 22, 2019:
http://mdb.cast.ru/mdb/4-2015/item3/article2

McDermott, Roger N., *Russia's Strategic Mobility: Supporting 'Hard Power' to 2020?* Stockholm, Sweden: FOI, FOI-R-3587-SE, April 2013. As of July 22, 2019:
https://www.foi.se/rest-api/report/FOI-R--3587--SE

Oboznik, "Задачи трубопроводных войск на территории Сирии" ["Tasks of the Pipeline Troops in Syria"], undated. As of July 22, 2019:
http://www.oboznik.ru/?p=52434#more-52434

Parraga, Marianna, and Alexandra Ulmer, "Special Report: Vladimir's Venezuela—Leveraging Loans to Caracas, Moscow Snaps Up Oil Assets," Reuters, August 11, 2017. As of July 22, 2019:
http://www.reuters.com/article/us-venezuela-russia-oil-specialreport/special-report-vladimirs-venezuela-leveraging-loans-to-caracas-moscow-snaps-up-oil-assets-idUSKBN1AR14U

Privalov, Aleksander, "The Russian Military Automotive Fleet," *Moscow Defense Brief*, No. 2 (34), 2013.

Ramm, Aleksey, and Lavrov, Anton, "Возмездие без дозаправки: Операция в Сирии показала слабые места Воздушно-космических сил" ["Retribution Without Refueling: Operation in Syria Reveals Weak Points in Aerospace Forces"], *Voyenno-Promyshlennyy Kuryer*, April 5, 2016. As of July 22, 2019: http://www.vpk-news.ru/articles/30078

Russian Defense Policy, "New Industrial-Logistical Complex," July 28, 2017. As of July 22, 2019: https://russiandefpolicy.blog/2017/07/28/new-industrial-logistic-complex

Russian Federation Presidential Edict 683, *Russian Federation's National Security Strategy*, December 31, 2015.

Russian Ministry of Defence, "Land Forces," webpage, undated. As of July 22, 2019: http://eng.mil.ru/en/structure/forces/ground.htm

———, "The Military Doctrine of the Russian Federation," December 25, 2014. As of July 22, 2019: http://rusemb.org.uk/press/2029

———, "Russian Military Medics Came Back to Russia from Palmyra," June 10, 2016. As of July 22, 2019: http://eng.mil.ru/en/news_page/country/more.htm?id=12087041@egNews

Southfront, "How the Russian Navy Provides 'Syrian Express,'" translated from Russian, December 28, 2015. As of July 22, 2019:
https://southfront.org/how-the-russian-navy-provides-syrian-transit

Spaulding, Hugo, Christopher Kozak, Christopher Harmer, Daniel Urchick, Jessica Lewis McFate, Jennifer Cafarella, Harleen Gambhir, and Kimberly Kagan, "Russian Deployment to Syria: Putin's Middle East Game Changer," Institute for the Study of War, September 17, 2015. As of July 22, 2019:
http://www.understandingwar.org/backgrounder/russian-deployment-syria-putin's-middle-east-game-changer

Torres, Patricia, and Nicholas Casey, "Armed Civilian Bands in Venezuela Prop Up Unpopular President," *New York Times*, April 22, 2017.

Tremaria, Stiven, "Violent Caracas: Understanding Violence and Homicide in Contemporary Venezuela," *International Journal of Conflict and Violence,* Vol. 10, No. 1, 2016, pp 61–76.

Turebekova, Aiman, "Kazakhstan, China Strengthen Defence Cooperation," *Astana Times,* October 16, 2015. As of July 22, 2019:
http://astanatimes.com/2015/10/kazakhstan-china-strengthen-defence-cooperation

Ulmer, Alexandra, and Marianna Parraga, "Exclusive: Russia, Venezuela Discuss Citgo Collateral Deal to Avoid U.S. Sanctions—Sources," Reuters, July 20, 2017. As of July 22, 2019:
https://www.reuters.com/article/us-oil-rosneft-citgo-exclusive/
exclusive-russia-venezuela-discuss-citgo-collateral-deal-to-avoid-u-s-sanctions-sources-idUSKBN1A52RN

"'Unprecedented Steps': Russian Military Explores Kuril Chain Island as Potential Pacific Fleet Base," RT, May 27, 2016. As of July 22, 2019:
https://www.rt.com/news/344539-kuril-island-russian-navy

Waldwyn, Tom, "Russian Military Lift Risks Atrophy," blog post, International Institute for Strategic Studies, July 6, 2017. As of July 22, 2019:
https://www.iiss.org/en/militarybalanceblog/blogsections/2017-edcc/july-c5e6/russian-military-a383

Werlau, Maria C., "Venezuela's Criminal Gangs: Warriors of Cultural Revolution," *World Affairs*, Vol 177, No. 2, July–August 2014, pp. 90–96.

Wintour, Patrick, "Russia Cancels Spanish Fuel Stop for Syria-Bound Warships," *The Guardian,* October 26, 2016. As of July 22, 2019:
https://www.theguardian.com/world/2016/oct/26/russia-cancels-spanish-fuel-stop-for-syria-bound-warships

Yang, Zi, "China's Private Security Companies: Domestic and International Roles," *China Brief,* Vol. 16, No. 15, October 4, 2016. As of July 22, 2019:
https://jamestown.org/program/chinas-private-security-companies-domestic-international-roles

Air Order of Battle

"Company Profile: Volga-Dnepr and the AN-124-100," *Moscow Defense Brief,* No. 1 (15), 2009.

Avia2, "Авиакомпания 224 Летный отряд (224 Flying Squad)," webpage, undated. As of July 22, 2019: http://avia2.ru/airlines/224-flying-squad

Barabanov, Mikhail, "Flight Unit 224: Russian MoD's Commercial Airline," *Moscow Defense Brief*, No. 3 (25), 2011. As of July 22, 2019:
https://mdb.cast.ru/mdb/3-2011/item4/article1

Felding, Daniel, "The Revival of Russia's Aviation Industry—New Russian Transport Aircraft," *Russia Insider*, March 10, 2016. As of July 22, 2019:
http://russia-insider.com/en/science-tech/revival-russias-aviation-industry-new-russian-transport-aircraft/ri13262

GlobalSecurity.org, "Airborne Assault Troops [VDV]—Airborne Equipment," webpage, last updated August 30, 2018a. As of December 20, 2017:
http://www.globalsecurity.org/military/world/russia/army-vdv-equipment.htm

———, "Military Transport Aviation (Voyennaya Transportnaya Aviatsiya—VTA)," webpage, last updated August 30, 2018b. As of July 22, 2019:
http://www.globalsecurity.org/military/world/russia/vta.htm

National Geospatial-Intelligence Agency, Digital Aeronautical Flight Information File, restricted-access database, data as of January 2017.

Nikolsky, Aleksey, "Russian Government's Non-Military Air Fleet: Structure and Procurement Policy," *Moscow Defense Brief*, No. 4 (36), 2013.

North Atlantic Treaty Organization, "Strategic Airlift Interim Solution (SALIS)," webpage, September 20, 2017. As of July 22, 2019:
http://www.nato.int/cps/en/natohq/topics_50106.htm

Prushinskiy, Aleksey, "Repair and Upgrade of Russian Non-Commercial Aircraft in 2011–2012," *Moscow Defense Brief*, No. 6 (44), 2014.

"Russia and the CIS," *Jane's Sentinel Security Assessment*, last updated February 23, 2017.

"Russia to Fill Gap Left by An-70 with More Il-76MD-90A Airlifters," *Jane's Defence Weekly*, January 13, 2016.

Russian Defense Policy, "The State of VTA," February 26, 2017. As of July 22, 2019:
https://russiandefpolicy.blog/2017/02/26/the-state-of-vta

"Ульяновский авиазавод построит в этом году три самолета Ил-76МД-90А" ["The Ulyanovsk Aircraft Plant Will Build Three Il-76MD-90A Aircraft this Year"], Interfax Military News Agency, February 28, 2017. As of July 22, 2019:
http://militarynews.ru/story.asp?rid=1&nid=443285

Volga-Dnepr, "An 124-100 Cargo Calculator," webpage, undated. As of July 22, 2019:
https://www.volga-dnepr.com/en/fleet/an-124/cc/#/en/an124

———, "IL-76 Cargo Calculator," webpage, undated. As July 22, 2019:
https://www.volga-dnepr.com/en/fleet/IL-76/cc/#/en/il76

Sea Order of Battle

Osborne, Andrew, "Russia Expands Military Transport Fleet to Move Troops Long Distances," Reuters, March 7, 2017. As of July 22, 2019:
http://www.reuters.com/article/russia-navy-expansion-idUSL5N1GK470

Parson, Joe, "How Russia Will Struggle to Keep Its Shipbuilders Afloat," Stratfor, January 20, 2016.

Polmar, Norman, and Michael Kofman, "New Russian Navy, Part 2: One Step Forward, Two Steps Back?" *Proceedings*, Vol. 143, No. 1, January 2017.

Voytenko, Mikhail, "Befriending a Devil," *FleetMon*, September 9, 2016. As of July 22, 2019:
https://www.fleetmon.com/maritime-news/2016/14799/befriending-devil

Rail Transport of Forces and Equipment

Aleksakov Y. F., "Перспективы длительного хранения техники и вооружения" ["Prospects for Long-Term Storage Equipment and Weapons"], Военная мысль [*Military Thought*], No. 1, January 2011, pp. 31–35.

Digital Forensic Research Lab, Atlantic Council, "Choo-Chooshka? New Russian Railway and Military Movement on the Ukrainian Border," *Medium*, August 15, 2017. As of July 22, 2019: https://medium.com/dfrlab/choo-chooshka-new-russian-railway-and-military-movement-on-the-ukrainian-border-39070aacc0ef

Coulibaly, Souleymane, Uwe Deichmann, William R. Dillinger, Marcel Ionescu-Heroiu, Ioannis N. Kessides, Charles Kunaka, and Daniel Saslavsky, *Eurasian Cities: New Realities Along the Silk Road*, Washington, D.C.: World Bank, 2012. As of July 22, 2019: http://documents.worldbank.org/curated/en/793131468256759023/Eurasian-cities-new-realities-along-the-silk-road

Globaltrans, "Industry Overview," webpage, last updated July 25, 2019. As of July 25, 2019: http://www.globaltrans.com/about-us/rail-industry-market/industry-overview

Goble, Paul, "The End of the Line for the Trans-Siberian Railroad?" *Eurasia Daily Monitor*, Vol. 13, No. 159, October 4, 2016. As of July 22, 2019: https://jamestown.org/program/end-line-trans-siberian-railroad

Kovalev Pavel, "Западные страшилки или мифы о российских вагонах" ["Western Horror Stories and Myths About Russian Military Trains"], Военно-политическое обозрение [*Military and Political Review*], August 10, 2017. As of July 22, 2019: http://www.belvpo.com/ru/85537.html

Lavrov, Anton, "Russian Railway Forces," *Moscow Defense Brief*, Vol. 3 (59), 2017

Marine Construction and Technologies, "Development of the Russian Sea Port Infrastructure. Automotive Logistics. Container Logistics in Russia," briefing slides, posted March 17, 2014. As of July 22, 2019: https://www.slideshare.net/alexandergoloviznin/ppt-goloviznin-201402

Murray, Bruce, *Russian Railway Reform Programme*, working paper, London: European Bank for Reconstruction and Development, June 2014.

Shepard, Wade, "2 Days from China to Europe by Rail? Russia Going for High-Speed Cargo Trains," *Forbes*, January 14, 2017. As of July 22, 2019: https://www.forbes.com/sites/wadeshepard/2017/01/14/2-days-from-china-to-europe-by-rail-russia-going-for-high-speed-cargo-trains/#5493a0de54af

Tolstov, G. S., "Проблемы и перспективы внедрения средств автоматической идентификации в интересах технического и тылового обеспечения ВС РФ" ["Problems and Prospects for the Introduction of Automatic Identification in the Interest of Technical and Logistic Support of the Armed Forces"], Военная мысль [*Military Thought*], No. 1, January 2010.

Historical Case Studies

Background and Context for Historical Deployments

Allard, Kenneth, "Soviet Airborne Forces and Preemptive Power Projection," *Parameters*, Vol. 10, No. 4, 1980, pp. 42–51.

Adomeit, Hannes, *Soviet Crisis Prevention and Management: Why and When Do the Soviet Leaders Take Risks?* Santa Monica, Calif.: RAND Corporation, OPS-008, October 1986. As of July 22, 2019: https://www.rand.org/pubs/occasional_papers-soviet/OPS008.html

Alexiev, Alexander R., *The New Soviet Strategy in the Third World*, Santa Monica, Calif.: RAND Corporation, N-1995-AF, June 1983. As of July 22, 2019: https://www.rand.org/pubs/notes/N1995.html

———, *The War in Afghanistan: Soviet Strategy and the State of the Resistance*, Santa Monica, Calif.: RAND Corporation, P-7038, November 1984. As of July 22, 2019:
https://www.rand.org/pubs/papers/P7038.html

Bar-Noi, Uri, "The Soviet Union and the Six-Day War: Revelations from the Polish Archives," Washington, D.C.: Wilson Center, July 7, 2011, https://www.wilsoncenter.org/publication/the-soviet-union-and-the-six-day-war-revelations-the-polish-archives

Bender, Gerald J., "The Eagle and the Bear in Angola," *Annals of the American Academy of Political and Social Science*, Vol. 489, No. 1, January 1987, pp. 123–132.

Benningsen, Alexandre, *The Soviet Union and Muslim Guerrilla Wars, 1920–1981: Lessons for Afghanistan*, Santa Monica, Calif.: RAND Corporation, N-1707/1, August 1981. As of July 22, 2019:
https://www.rand.org/pubs/notes/N1707z1.html

Bienen, Henry, "Perspectives on Soviet Intervention in Africa," *Political Science Quarterly*, Vol. 95, No. 1, Spring 1980, pp. 29–42.

Blumenstock, Elvis E., *A Look at Soviet Deep Operations: Is There an Amphibious Operational Maneuver Group in the Marine Corps' Future?* thesis, Quantico, Va.: U.S. Marine Corps Command and Staff College, 1994.

Bruckoff, Patricia A., Aleksei Georgievich Arbatov, Abraham S. Becker, Axel Leijonhufvud, and P. E. Anderson, *Russia and Her Neighbors: Symposium Report*, Santa Monica, Calif.: RAND Corporation, OPS-026, May 1992. As of July 22, 2019:
https://www.rand.org/pubs/occasional_papers-soviet/OPS026.html

Central Intelligence Agency, Director of Central Intelligence, *National Intelligence Estimate: Warsaw Pact Forces for Operations in Eurasia*, declassified intelligence analysis report, Langley, Va., September 1971. As of July 22, 2019:
https://www.cia.gov/library/readingroom/docs/1971-09-09.pdf

Central Intelligence Agency, Directorate of Intelligence, *The Czechoslovak-Soviet Struggle*, declassified intelligence analysis report, Langley, Va., July 12, 1968a. As of July 22, 2019:
https://www.cia.gov/library/readingroom/docs/CIA-RDP94T00754R000200290008-1.pdf

———, *Costs to Czechoslovakia, and to the Warsaw Pact Powers, of Actions Taken Against the Czechoslovak Economy*, declassified intelligence analysis report, Langley, Va., September 19, 1968b. As of July 22, 2019:
https://www.cia.gov/library/readingroom/docs/DOC_0000126873.pdf

———, *Military Costs of the Warsaw Pact Invasion of Czechoslovakia*, declassified intelligence analysis report, Langley, Va., September 19, 1968c. As of July 22, 2019:
https://www.cia.gov/library/readingroom/docs/DOC_0000126872.pdf

———, *Soviet Policy and the 1967 Arab-Israeli War*, declassified intelligence analysis report, March 16, 1970. As of July 22, 2019:
https://www.cia.gov/library/readingroom/docs/DOC_0001408643.pdf

———, "Organization of Soviet Foreign Military and Economic Aid," Washington, D.C., May 1976. As of July 22, 2019:
https://www.cia.gov/library/readingroom/docs/DOC_0000498599.pdf

———, "The Soviet Military Advisory and Training Program for the Third World," 1983a. As of July 22, 2019:
https://www.cia.gov/library/readingroom/docs/CIA-RDP08S01350R000100260001-8.pdf

———, *Ethiopia: The Impact of Soviet Military Assistance*, declassified intelligence analysis report, January 1983b. As of July 22, 2019:
https://www.cia.gov/library/readingroom/docs/DOC_0000496797.pdf

———, *Soviet Naval Activity Outside Home Waters During 1982*, declassified intelligence analysis report, Langley, Va., July 1983c. As of July 22, 2019:
https://www.cia.gov/library/readingroom/docs/DOC_0005173426.pdf

———, *The Soviet Invasion of Afghanistan: Five Years After*, declassified intelligence analysis report, May 1985. As of July 22, 2019:
https://www.cia.gov/library/readingroom/docs/DOC_0000496704.pdf

———, *Soviet Naval Activities Outside Home Waters in 1985*, declassified intelligence analysis report, Langley, Va., October 1986. As of July 22, 2019:
https://www.cia.gov/library/readingroom/docs/DOC_0000499548.pdf

Central Intelligence Agency, Directorate of Operations, *Military Thought (USSR): A General Review of Modern Military Doctrine*, declassified intelligence analysis report, Langley, Va., March 7, 1974a. As of July 22, 2019:
https://www.cia.gov/library/readingroom/docs/DOC_0001199073.pdf

———, *Military Thought (USSR): Surprise in Starting a War*, declassified intelligence analysis report, Langley, Va., June 27, 1974b. As of July 22, 2019:
https://www.cia.gov/library/readingroom/docs/CIA-RDP10-00105R000100710001-7.pdf

———, *Military Thought (USSR): The Preparation and Conduct of an Operation by the Armed Forces in a Theater of Military Operations in the Initial Period of War*, declassified intelligence analysis report, Langley, Va., May 10, 1976. As of July 22, 2019:
https://www.cia.gov/library/readingroom/docs/DOC_0001199099.pdf

———, *Warsaw Pact Journal: From the Experience of the Actions of Rear Services Units and Facilities in the TRANZIT-74 Exercise*, declassified intelligence analysis report, Langley, Va., September 14, 1977.

———, *USSR General Staff Academy Lessons: Study and Critique of the Plan of the Amphibious Landing Operation*, declassified intelligence analysis report, Langley, Va., 1980. As of July 22, 2019:
https://www.cia.gov/library/readingroom/docs/DOC_0001197552.pdf

Central Intelligence Agency, Historical Collections Division, *Strategic Warning and the Role of Intelligence: Lessons Learned from the 1968 Soviet Invasion of Czechoslovakia*, Washington, D.C., 2013. As of July 22, 2019:
https://www.cia.gov/library/publications/cold-war/czech-invasion/soviet%20-czech-invasion.pdf

Central Intelligence Agency, National Foreign Assessment Center, *Soviet Amphibious Forces: Tasks and Capabilities in General War and Peacetime*, declassified intelligence analysis report, Langley, Va., 1979.

Central Intelligence Agency, Office of National Estimates, *The Crisis in Czechoslovakia*, declassified intelligence analysis report, Langley, Va., July 12, 1968.

Central Intelligence Agency, Office of Strategic Research, *Status of Soviet Strategic Offensive Forces 1 February 1975*, declassified intelligence analysis report, Langley, Va., 1975. As of July 22, 2019:
https://www.cia.gov/library/readingroom/docs/DOC_0005672884.pdf

Central Intelligence Agency, Office of Current Intelligence, *Soviet Military Forces in Cuba*, declassified intelligence analysis report, Current Intelligence Weekly Review, c. January 1, 1962. As of July 22, 2019:
https://www.cia.gov/library/readingroom/docs/DOC_0001161979.pdf

———, "Soviet Forces in Cuba," declassified memorandum, May 7, 1963. As of July 22, 2019:
https://www.cia.gov/library/readingroom/docs/CIA-RDP79T00429A001100040005-6.pdf

———, "The Soviet Presence in Yemen," memorandum, May 7, 1963. As of July 22, 2019:
https://www.cia.gov/library/readingroom/docs/CIA-RDP79T00429A001100040006-5.pdf

———, "Relations Between Syria and the USSR," memorandum, June 1, 1976. As of July 22, 2019:
https://www.cia.gov/library/readingroom/docs/CIA-RDP85T00353R000100290001-4.pdf

Crutcher, Michael H., ed., *The Russian Armed Forces at the Dawn of the Millennium*, conference proceedings, Carlisle, Pa.: U.S. Army War College, December 2000. As of July 22, 2019:
https://apps.dtic.mil/dtic/tr/fulltext/u2/a423593.pdf

Cunningham, Erin, and Karen DeYoung, "Strikes from Iranian Air Base Show Russia's Expanding Footprint in the Middle East," *Washington Post*, August 2016.

Defense Intelligence Agency, *Soviet Electronic Countermeasures During Invasion of Czechoslovakia*, declassified intelligence report, Washington, D.C., 1968.

———, *Cuba: Soviet Military Activities*, declassified intelligence report, Washington, D.C., September 26, 1978. As of July 22, 2019:
https://www.cia.gov/library/readingroom/docs/CIA-RDP06T01849R000100030037-2.pdf

———, "Potential for Soviet Intervention in Syria," declassified memorandum, May 13, 1980. As of July 22, 2019:
https://www.cia.gov/library/readingroom/docs/CIA-RDP83B01027R000300170009-1.pdf

Dismukes, Bradford, and James M. McConnell, eds., *Soviet Naval Diplomacy*, Elmsford, N.Y.: Pergamon Press, 1979.

Dowling, Timothy C., ed., *Russia at War: From the Mongol Conquest to Afghanistan, Chechnya, and Beyond*, Santa Barbara, Calif.: ABC-CLIO, 2015.

Dragnich, George S., *The Soviet Union's Quest for Access to Naval Facilities in Egypt Prior to the June War of 1967*, AD-786, Arlington, Va.: Center for Naval Analyses, July 1974.

Dzirkals, Lilita I., *"Lightning War" in Manchuria: Soviet Military Analysis of the 1945 Far East Campaign*, Santa Monica, Calif.: RAND Corporation, P-5589, January 1976. As of July 22, 2019:
https://www.rand.org/pubs/papers/P5589.html

Edmonds, Martin, and John Skitt, "Current Soviet Maritime Strategy and NATO," *International Affairs*, Vol. 45, No. 1, 1969, pp. 28–43.

Felgenhauer, Pavel, "Russia's Secret Operations," *Perspective*, Vol. 12, No. 1, September–October 2001.

Fukuyama, Francis, *The Soviet Threat to the Persian Gulf*, Santa Monica, Calif.: RAND Corporation, P-6596, March 1981. As of July 22, 2019:
https://www.rand.org/pubs/papers/P6596.html

———, *Soviet Civil-Military Relations and the Power Projection Mission*, Santa Monica, Calif.: RAND Corporation, R-3504-AF, April 1987. As of July 22, 2019:
https://www.rand.org/pubs/reports/R3504.html

Fukuyama, Francis, Scott Bruckner, and Sally W. Stoecker, *Soviet Political Perspectives on Power Projection*, Santa Monica, Calif.: RAND Corporation, N-2430-A, March 1987. As of July 22, 2019:
https://www.rand.org/pubs/notes/N2430.html

Galeotti, Mark, *Russia's Wars in Chechnya 1994–2009*, New York: Osprey, 2014.

Garthoff, Raymond L., "Estimating Soviet Military Force Levels: Some Light from the Past," *International Security*, Vol. 14, No. 4, Spring 1990, pp. 93–116.

Gelman, Harry, *The Soviet Military Leadership and the Question of Soviet Deployment Retreats*, Santa Monica, Calif.: RAND Corporation, R-3664-AF, November 1988. As of July 22, 2019:
https://www.rand.org/pubs/reports/R3664.html

Giles, Keir, *Russia's "New" Tools for Confronting the West: Continuity and Innovation in Moscow's Exercise of Power*, London: Chatham House, March 2016. As of July 22, 2019:
https://www.chathamhouse.org/sites/default/files/publications/2016-03-russia-new-tools-giles.pdf

Ginor, Isabella, "The Russians Were Coming: The Soviet Military Threat in the 1967 Six-Day War," *Middle East Review of International Affairs*, Vol. 4, No. 4, December 2000, pp. 44–59.

Glantz, David M., *The Soviet Airborne Experience*, Fort Leavenworth, Kan.: Combat Studies Institute, U.S. Army Command and General Staff College, November 1984.

Goldstein, Lyle J., and Yuri M. Zhukov, "A Tale of Two Fleets—A Russian Perspective on the 1973 Naval Standoff in the Mediterranean," *Naval War College Review*, Vol. 57, No. 2, Spring 2004, Article 4.

Golts, Alexander, and Michael Kofman, *Russia's Military: Assessment, Strategy, and Threat*, Washington, D.C.: Center on Global Interests, June 2016. As of July 22, 2019:
http://globalinterests.org/wp-content/uploads/2016/06/Russias-Military-Center-on-Global-Interests-2016.pdf

Higham, Robin, and Frederick W. Kagan, ed., *The Military History of the Soviet Union*, New York: Palgrave, 2002.

Holcomb, Jr., James F., *Soviet Airborne Forces and the Central Region: Problems and Perceptions*, Fort Leavenworth, Kan.: U.S. Army Combined Arms Center, 1987. As of July 22, 2019:
https://apps.dtic.mil/dtic/tr/fulltext/u2/a193561.pdf

Hosmer, Stephen T., and Thomas W. Wolfe, *Soviet Policy and Practice Towards Third World Conflicts*, Lexington, Mass.: Lexington Books, 1982.

Klinghoffer, Arthur J., *The Soviet Union and Angola*, Carlisle, Pa.: Strategic Studies Institute, U.S. Army War College, May 10, 1980. As of July 22, 2019:
https://apps.dtic.mil/dtic/tr/fulltext/u2/a088004.pdf

Lambeth, Benjamin S., *How to Think About Soviet Military Doctrine*, Santa Monica, Calif.: RAND Corporation, P-5939, February 1978. As of July 22, 2019:
https://www.rand.org/pubs/papers/P5939.html

Nechepurenko, Ivan, "Russia Seeks to Reopen Military Bases in Vietnam and Cuba," *New York Times*, October 7, 2016.

Oliker, Olga, *Soft Power, Hard Power, and Counterinsurgency: The Early Soviet Experience in Central Asia and Its Implications*, Santa Monica, Calif.: RAND Corporation, WR-547-RC, February 2008. As of July 22, 2019:
https://www.rand.org/pubs/working_papers/WR547.html

Oliker, Olga, and Tanya Charlick-Paley, *Assessing Russia's Decline: Trends and Implications for the United States and the U.S. Air Force*, Santa Monica, Calif.: RAND Corporation, MR-1442-AF, 2002. As of July 22, 2019:
https://www.rand.org/pubs/monograph_reports/MR1442.html

Pikhoia, R. G., "Czechoslovakia in 1968: A View from Moscow According to Central Committee Documents," *Russian Studies in History*, Vol. 44, No. 3, Winter 2005–2006, pp. 35–80.

Quandt, William B., *Algerian Military Development: The Professionalization of a Guerrilla Army*, Santa Monica, Calif.: RAND Corporation, P-4792, March 1972. As of July 22, 2019:
https://www.rand.org/pubs/papers/P4792.html

Ra'anan, Gavriel D., *The Evolution of the Soviet Use of Surrogates in Military Relations with the Third World, with Particular Emphasis on Cuban Participation in Africa*, Santa Monica, Calif.: RAND Corporation, P-6420, December 1979. As of July 22, 2019:
https://www.rand.org/pubs/papers/P6420.html

"Report from Soviet Deputy Minister of Internal Affairs Perevertkin on the Movement of Soviet Troops into Hungary," translation, October 24, 1956. As of July 22, 2019:
http://digitalarchive.wilsoncenter.org/document/111967

Ross, Dennis, "Considering Soviet Threats to the Persian Gulf," *International Security*, Vol. 6, No. 2, Fall 1981, pp. 159–180.

"Russia Used Iranian Airfield for Syrian Operation at Tehran's Invitation—Official," RT, August 23, 2016. As of July 22, 2019:
https://www.rt.com/news/356888-russia-iran-airfield-syria

Sadykiewicz, Michael, *Organizing for Coalition Warfare: The Role of East European Warsaw Pact Forces in Soviet Military Planning*, Santa Monica, Calif.: RAND Corporation, R-3559-RC, September 1988. As of July 22, 2019:
https://www.rand.org/pubs/reports/R3559.html

Schmid, Alex P., *Soviet Military Interventions Since 1945*, New Brunswick, N.J.: Transaction Books, 1985.

Shlapak, David A., and Michael Johnson, *Reinforcing Deterrence on NATO's Eastern Flank: Wargaming the Defense of the Baltics*, Santa Monica, Calif.: RAND Corporation, RR-1253-A, 2016. As of July 22, 2019:
https://www.rand.org/pubs/research_reports/RR1253.html

Stern, Ellen P., ed., *The Limits of Military Intervention*, Beverly Hills, Calif.: Sage Publications, 1977.

Stevens, Christopher, "The Soviet Union and Angola," *African Affairs*, Vol. 75, No. 299, April 1976, pp. 137–151.

Stewart, Richard A., "Soviet Military Intervention in Iran, 1920–1946," *Parameters*, Vol. 11, No. 4, 1981, pp. 24–33.

Stone, David R., *A Military History of Russia: From Ivan the Terrible to the War in Chechnya*, Westport, Conn.: Praeger, 2006.

Thomas, Timothy L., *Russia Military Strategy: Impacting 21st Century Reform and Geopolitics*, Ft. Leavenworth, Kan.: Foreign Military Studies Office, U.S. Army Combined Arms Center, 2015.

Thompson, W. Scott, *The Projection of Soviet Power*, Santa Monica, Calif.: RAND Corporation, P-5988, 1977.

Trapans, Andris, *Logistics in Recent Soviet Military Writings*, Santa Monica, Calif.: RAND Corporation, RM-5062-PR, August 1966. As of July 22, 2019:
https://www.rand.org/pubs/research_memoranda/RM5062.html

Turbiville, Jr., Graham H., "Soviet Airborne Operations in Theater War," *Foreign Policy*, Vol. 23, Nos. 1–2, 1986, pp. 160–183.

Ürményházi, Attila J., *The Hungarian Revolution-Uprising, Budapest 1956*, Washington, D.C.: Hungarian-American Federation, 2006. As of July 22, 2019:
http://www.americanhungarianfederation.org/docs/Urmenyhazi_HungarianRevolution_1956.pdf

U.S. Department of Defense, *Soviet Military Power, 1986*, Washington, D.C.: U.S. Government Printing Office, 1986.

Valenta, Jiri, "From Prague to Kabul: The Soviet Style of Invasion," *International Security*, Vol. 5, No. 2, Fall 1980, pp. 114–141.

Van Oudenaren, John, *The Soviet Union and Second-Area Actions*, Santa Monica, Calif.: RAND Corporation, N-2035-FF/RC, September 1983. As of July 22, 2019:
https://www.rand.org/pubs/notes/N2035.html

West, Lowry A., *Soviet Airborne Operations*, Garmisch, Germany: U.S. Army Russian Institute, 1980. As of July 22, 2019:
https://apps.dtic.mil/dtic/tr/fulltext/u2/a115148.pdf

Winterford, David, *Assessing the Soviet Naval Build-Up in Southeast Asia: Threats to Regional Security*, Monterey, Calif.: Naval Postgraduate School, September 1988. As of July 22, 2019:
https://apps.dtic.mil/dtic/tr/fulltext/u2/a201291.pdf

Wintour, Patrick, and Shaun Walker, "Vladimir Putin Orders Russian Forces to Begin Withdrawal from Syria," *The Guardian*, March 15, 2016. As of July 22, 2019:
https://www.theguardian.com/world/2016/mar/14/vladimir-putin-orders-withdrawal-russian-troops-syria

Wolfe, Thomas W., *The Soviet Quest for More Globally Mobile Military Powers*, Santa Monica, Calif.: RAND Corporation, RM-5554-PR, December 1967. As of July 22, 2019:
https://www.rand.org/pubs/research_memoranda/RM5554.html

———, *The USSR and the Arab East*, Santa Monica, Calif.: RAND Corporation, P-4194, September 1969. As of July 22, 2019:
https://www.rand.org/pubs/papers/P4194.html

Zagoria, Donald S., "The USSR and Asia in 1984," *Asian Survey*, Vol. 25, No. 1, January 1985, pp. 21–32.

Historical Case Study: Syrian Civil War

Barabanov, M., and A. Vasiliev, Сирийский рубеж [*Syrian Frontier*], Moscow: Centre for Analysis of Strategies and Technologies, 2016. As of July 22, 2019:
http://cast.ru/upload/iblock/686/6864bf9d4485b9cd83cc3614575e646a.pdf

Baranets, Victor, "Генерал-полковник Андрей Картаполов: У России может появиться база в Сирии" ["Colonel-General Andrei Kartapolov: Russia May Have a Base in Syria"], *Komsomolskaya Pravda*, October 16, 2015. As of July 22, 2019:
https://www.kp.ru/daily/26446/3316981

Binnie, Jeremy, "Iskaner Missile Launcher Spotted in Syria," *Jane's Defence Weekly*, March 31, 2016.

Cenciotti, David, "Online Flight Tracking Provides Interesting Details About Russian Air Bridge to Syria," *The Aviationist*, September 11, 2015a. As of July 22, 2019:
https://theaviationist.com/2015/09/11/ads-b-exposes-russian-air-bridge-to-syria

———, "Here's How the Russian Air Force Moved 28 Aircraft to Syria (Almost) Undetected," *The Aviationist*, September 23, 2015b. As of July 22, 2019:
https://theaviationist.com/2015/09/23/how-the-russians-deployed-28-aircraft-to-syria

Gavrilov, Yuri, "Сирия: русский гром" ["Syria: Russian Thunder"], transcript of interview with commander of Russian forces in Syria, *Rossiyskaya Gazeta*, March 23, 2016. As of July 22, 2019:
https://rg.ru/2016/03/23/aleksandr-dvornikov-dejstviia-rf-v-korne-perelomili-situaciiu-v-sirii.html
English translation, as of July 22, 2019:
http://csef.ru/en/oborona-i-bezopasnost/423/komanduyushhij-gruppirovkoj-vojsk-rf-v-sirii-dal-pervoe-intervyu-rossijskoj-gazete-6644

Institute for the Study of War, "Significant Offensives in Syria: June 6–July 9, 2015," July 9, 2015. As of July 22, 2019:
http://www.understandingwar.org/backgrounder/significant-offensives-syria-june-6-june-9-2015

"Islamic State and the Crisis in Iraq and Syria in Maps," BBC News, November 2, 2016. As of July 22, 2019:
http://www.bbc.com/news/world-middle-east-27838034

Kozak, Christopher, "Posture of Syrian Regime and Allies, September 14, 2015," Institute for the Study of War, September 14, 2015a. As of July 22, 2019:
http://www.understandingwar.org/backgrounder/posture-syrian-regime-and-allies-september-14-2015

———, "Syria 90-Day Strategic Forecast: The Regime and Allies," Institute for the Study of War, September 22, 2015b. As of July 22, 2019:
http://www.understandingwar.org/backgrounder/syria-90-day-strategic-forecast-regime-and-allies

Maritime Executive, "Russia Moves Forward with Syrian Naval Base," December 13, 2017. As of July 22, 2019:
https://maritime-executive.com/article/russia-moves-forward-with-syrian-naval-base

National Antiterrorism Committee, "Выступление первого заместителя руководителя аппарата НАК Е.П. Ильина на Всероссийском семинаре-совещании руководителей патриотических объединений" ["Speech of the First Deputy Head of the National Antiterrorism Committee Apparatus E. P. Ilyin at the All-Russian Seminar Meeting of Heads of Patriotic Associations"], October 27, 2015. As of July 22, 2019:
http://nac.gov.ru/publikacii/vystupleniya-i-intervyu/vystuplenie-pervogo-zamestitelya.html

Ramm, Aleksey, and Lavrov, Anton, "Возмездие без дозаправки: Операция в Сирии показала слабые места Воздушно-космических сил" ["Retribution Without Refueling: Operation in Syria Reveals Weak Points in Aerospace Forces"], *Voyenno-Promyshlennyy Kuryer*, April 5, 2016. As of July 22, 2019:
http://www.vpk-news.ru/articles/30078

"Russia Bypasses NATO's Aerial Blockade, Renews Airlift to Syria Via Iran," *OSNet Daily*, September 9, 2015. As of November 5, 2016. As of July 22, 2019:
http://osnetdaily.com/2015/09/russia-bypasses-natos-aerial-blockade-renews-airlift-to-syria-via-iran

Zain, H., "Russian Defense Ministry: Russian Airstrikes Kill More Than 30 Terrorists, Including Commanders," Syrian Arab News Agency, November 17, 2016. As of July 22, 2019:
https://sana.sy/en/?p=93621

Spaulding, Hugo, Christopher Kozak, Christopher Harmer, Daniel Urchick, Jessica Lewis McFate, Jennifer Cafarella, Harleen Gambhir, and Kimberly Kagan, "Russian Deployment to Syria: Putin's Middle East Game Changer," Institute for the Study of War, September 17, 2015. As of July 22, 2019:
http://www.understandingwar.org/backgrounder/russian-deployment-syria-putin's-middle-east-game-changer

Sutyagin, Igor, "Detailing Russian Forces in Syria," RUSI, November 13, 2015. As of July 22, 2019:
https://rusi.org/publication/rusi-defence-systems/detailing-russian-forces-syria

"Syria: The Story of the Conflict," BBC News, March 11, 2016. As of July 22, 2019:
http://www.bbc.com/news/world-middle-east-26116868

"Syria Civil War Timeline: A Summary of Critical Events," Deutsche Welle, August 14, 2017. As of July 22, 2019:
http://www.dw.com/en/syria-civil-war-timeline-a-summary-of-critical-events/a-40001379

Historical Case Study: Soviet-Afghan War

Baumann, Robert, *Russian-Soviet Unconventional Wars in the Caucasus, Central Asia, and Afghanistan*, Fort Leavenworth, Kan.: Combat Studies Institute, U.S. Army Command and General Staff College, Leavenworth Papers No. 20, 1993. As of July 22, 2019:
https://history.army.mil/html/books/107/107-1

Central Intelligence Agency, Director of Central Intelligence, "The Soviet Invasion of Afghanistan: Implications for Warning," declassified memorandum, October 1980. As of July 22, 2019:
https://www.cia.gov/library/readingroom/docs/DOC_0000278538.pdf

———, "A Review of Intelligence Performance in Afghanistan," declassified memorandum, April 9, 1984. As of July 22, 2019:
https://www.cia.gov/library/readingroom/docs/CIA-RDP86B00269R001100100003-5.pdf

Grau, Les, ed., *The Bear Went over the Mountain: Soviet Combat Tactics in Afghanistan*, Washington, D.C.: National Defense University Press, 2005. As of July 22, 2019:
https://apps.dtic.mil/dtic/tr/fulltext/u2/a316729.pdf

Grau, Les, and William Jorgensen, "Medical Support in a Counter-Guerrilla War: Epidemiologic Lessons Learned in the Soviet-Afghan War," *U.S. Army Medical Department Journal*, May–June 1995. As of July 22, 2019:
https://community.apan.org/wg/tradoc-g2/fmso/m/fmso-monographs/240304

Lyakhovskiy, Aleksandr Antonovich, *Inside the Soviet Invasion of Afghanistan and the Seizure of Kabul, December 1979*, Washington, D.C.: Wilson Center, Working Paper No. 51, 2007. As of July 22, 2019:
https://www.wilsoncenter.org/publication/inside-the-soviet-invasion-afghanistan-and-the-seizure-kabul-december-1979

MacEachin, Douglas, "Predicting the Soviet Invasion of Afghanistan: The Intelligence Community's Record," Washington, D.C.: Central Intelligence Agency, Center for the Study of Intelligence, last updated June 28, 2008. As of July 22, 2019:
https://www.cia.gov/library/center-for-the-study-of-intelligence/csi-publications/books-and-monographs/predicting-the-soviet-invasion-of-afghanistan-the-intelligence-communitys-record/predicting-the-soviet-invasion-of-afghanistan-the-intelligence-communitys-record.html

Russian General Staff, *The Soviet-Afghan War: How a Superpower Fought and Lost*, Lester Grau and Michael Gress, trans. and ed., Lawrence, Kan.: University Press of Kansas, 2002.

Stone, David R., *A Military History of Russia: From Ivan the Terrible to the War in Chechnya,* Westport, Conn.: Praeger, 2006.

Turbiville, Graham H., Jr., "Ambush! The Road War in Afghanistan," *Army*, January 1988. As of July 22, 2019:
https://community.apan.org/wg/tradoc-g2/fmso/m/fmso-monographs/240945